A MYSTERIOUS MASTERPIECE.
THE WORLD OF THE LINDER GALLERY

| A MYSTERIOUS MASTERPIECE.

THE WORLD OF THE LINDER GALLERY

edited by
Michael John Gorman

Introduction
Michael John Gorman
James M. Bradburne

| ALIAS

This publication
was written to coincide
with the exhibition

GALILEO
IMAGES OF THE UNIVERSE
FROM ANTIQUITY
TO THE TELESCOPE

Florence
Palazzo Strozzi
13 March
30 August
2009

A PUBLICATION OF
Fondazione Palazzo Strozzi

CONCEPT AND CREATIVE
DIRECTION
James M. Bradburne

EDITED BY
Michael John Gorman

TEXTS
Michael John Gorman
James M. Bradburne
Alexander Marr

PARTICIPANTS
IN THE CONVERSATION
James M. Bradburne
Ron e Barbara Cordover
Alexander Marr
Michael John Gorman
Pamela Smith
Lawrence Weschler

NEW YORK EVENT
COORDINATION
Stephanie Steiker

TRANSCRIPTION
Daniel Erker

EDITING
Andrea Paoletti
Monica Fintoni

EDITORIAL
COORDINATION
Ludovica Sebregondi

GRAPHIC DESIGN
AND LAYOUT
RovaiWeber design

ACKNOWLEDGMENTS
Special thanks to Ron
and Barbara Cordover.
Thanks also to
Miel de Botton-Aynsley.

PHOTOGRAPHIC CREDITS

© Blauel/Gnamm/Artothek (p. 121)
© PhotoserviceElecta/AKG Images (p. 19)
© Skokloster Castle. All rights reserved. Photo Samuel Uhrdin (p. 11)
© The National Gallery, London (p. 89)
© Vincent Desiderio, courtesy Marlborough Gallery, New York (p. 59)
Biblioteca Marucelliana, Florence (p. 32)*
Biblioteca Nazionale Centrale di Firenze, Florence (pp. 29, 30, 31, 33, 35, 37, 68, 109, 116)*
Biblioteca Riccardiana, Florence (p. 25)*
Collezione Subert, courtesy Michele Subert, Milan (p. 115)
Christie's London (p. 108)
Courtesy of The Trustees of the National Library of Scotland (p. 116)
Photo Rabatti & Domingie Photography, Florence (pp. 6–7, 19, 23, 30, 46–7, 49, 50, 51, 53, 60–1, 63, 65, 67, 70–1, 72, 74, 76–7, 78, 79, 80–1, 84, 85, 86, 87, 91, 92–3, 94, 97, 98, 100, 103, 104, 107, 112–3, 114, 117, 127–8, 129)
Photo Tamara Schlesinger, New York (pp. 42, 44)
Galleria degli Uffizi, Photo Paolo Tosi, Florence (p. 119)*
Museo Nacional del Prado, Madrid, Photographic Archive (pp. 16–7)
The Royal Collection © 2009 Her Majesty Queen Elizabeth II (pp. 20–1, 63, 72, 82, 84, 85, 119)

* With the permission of the Italian Ministry for Cultural Affairs. Further reproduction in any form is strictly prohibited.

www.palazzostrozzi.org

Printed in Italy
by Lito Terrazzi, Florence

June 2009

ISBN 978-88-96532-02-7

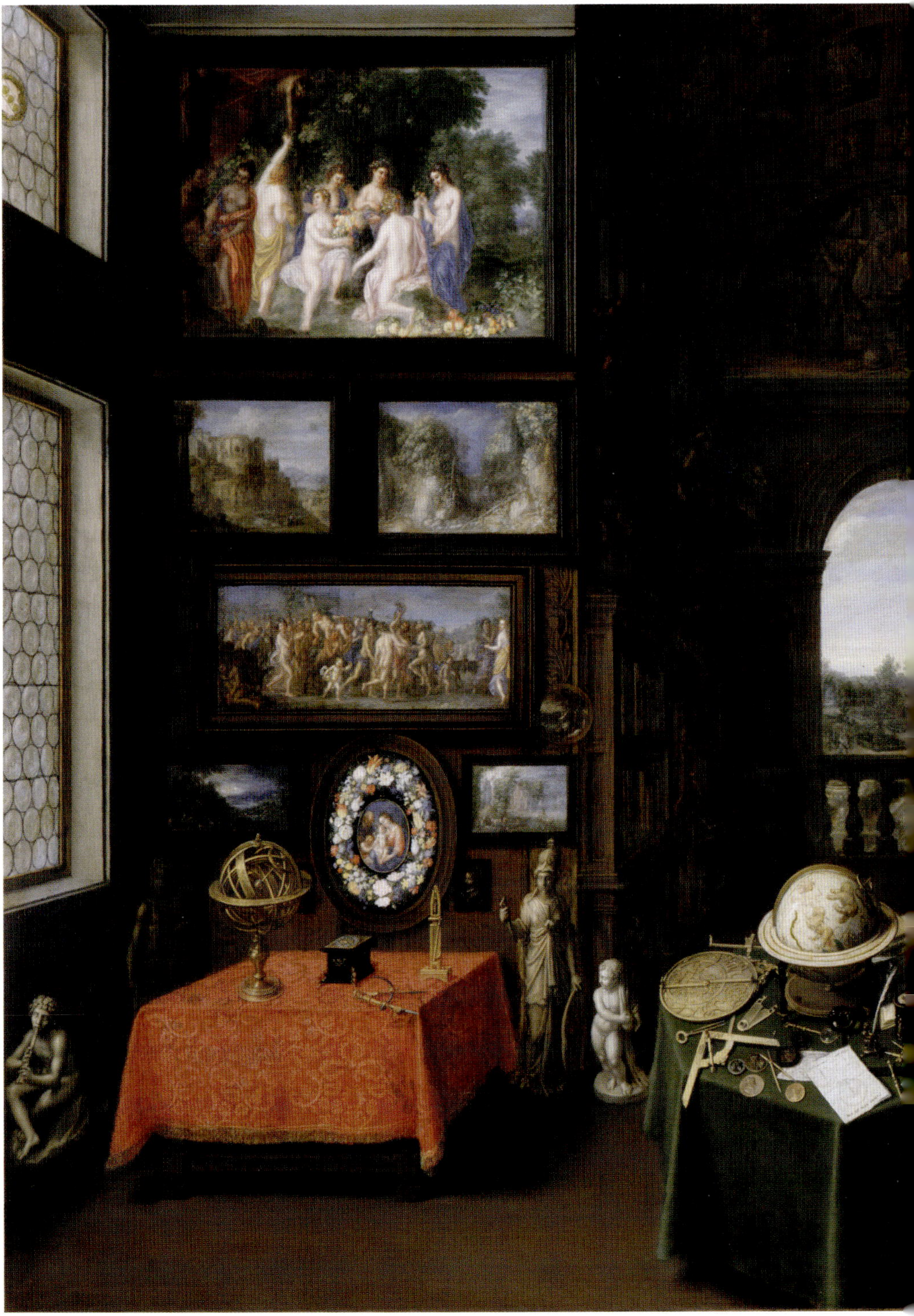

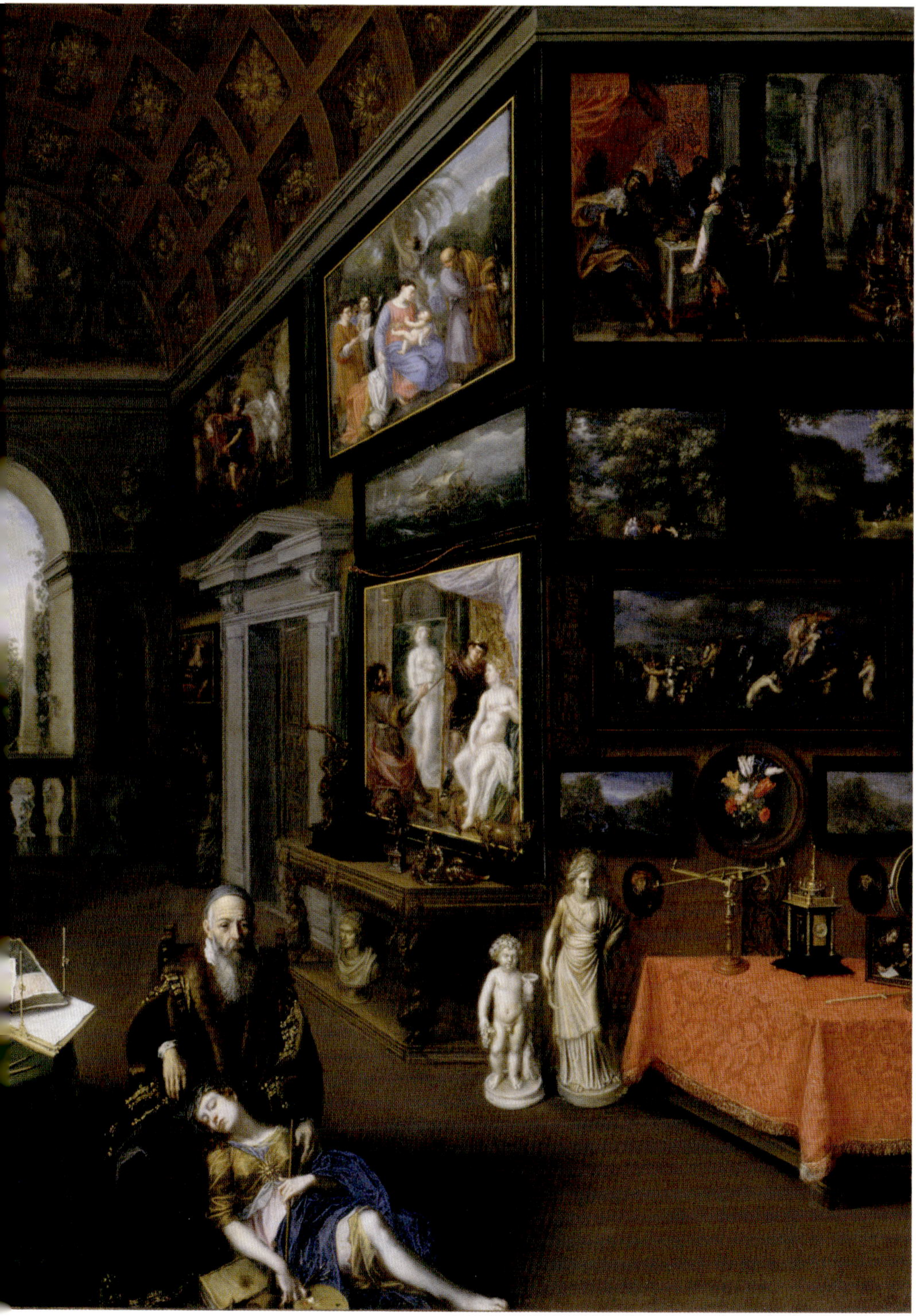

THE WORLD OF THE LINDER GALLERY

This book is about one painting—an extraordinary painting. Extraordinary in its detail, extraordinary in its artistic quality and extraordinary because until now it has been virtually unknown. What the painting actually shows is relatively obvious. The painting seems to belong to a world of artists, scientists and collectors. However, there are many subtle and mysterious references within it, not all of which are presently understood. How did this interior come to be painted? Why? For whom and by whom? To show what? Everything about the painting is a mystery, a puzzle. Like all mysteries, there are clues, if you know how to read them. Like all puzzles, however difficult, there are solutions. This book is based on a conversation about this painting among specialists and generalists alike trying to unravel these clues. However, the puzzle is not completely solved, and this book is an invitation to join in the discussion not only about a single painting, but also about a unique moment in the history of ideas, a moment when Europe stood at the threshold of a new era of observational astronomy, marked by Galileo's use of the telescope to observe the moon and planets.

The challenge is to see the world in terms of the choices faced by the historical actor. Too often we see history with an all-embracing hindsight that comes from already knowing the outcome of past events. It is often more useful—and indeed more interesting—to try to understand how our predecessors behaved when faced with a limited range of possible decisions, ignorant of their outcome, and indeed of all the outcomes that have since become our shared history. As Alan Bennett has the art historian Anthony Blunt say in *A Question of Attribution*, "Michelangelo didn't know Raphael would come along and do it better". Our own ability to see into the future serves as a reminder—it is as flawed and partial as that of any historical figure we try to understand. As events recede in time, much of the context is lost and the rich fog of considerations that shapes our own thinking—the television, the newspapers and the Internet, our world view coloured by Freud, Marx, Einstein, Madonna and Britney Spears—eventually evaporates, leaving a crude and often frustratingly inadequate residue of documents such as texts, objects and paintings.

Paintings in the sixteenth and seventeenth centuries were not just decoration, as some might understand them today. What can we say about how images were used in the late Renaissance and Baroque courts? Power in the sixteenth and early seventeenth centuries was mediated by symbols. Lacking the means of communication that make the contemporary world a buzzing confusion of competing images, those who commissioned paintings made sure that every project they undertook was laden with imagery meant to communicate to both the literate and illiterate onlooker. The Habsburg emperors provide a good example of the Renaissance obsession with image-making. From the massive statues of Hercules outside the Prague Castle gates that proclaimed Habsburg supremacy at the same time as alerting viewers to their putatively imperial lineage, to the double-headed Habsburg eagle that spread its all-encompassing wings over the continent, to the mystical device *AEIOU* used by Frederick III (1415–93) that stood for *Alles Erdreich Ist Österreich Untertan* ('Everything earthly is subject to Austria'), the Habsburgs made sure that their message was ubiquitous and unambiguous—they were in power and meant to keep it.

In addition to representing power, images were often believed to have another, more instrumental value—they could draw down astral influences. To understand the Renaissance belief in these occult forces, one of the key figures is Marsilio Ficino (1433–99). Following the lead of George Gemistus Plethon (*c.* 1355–1450/2), Ficino tried to conflate Christianity and Platonism, while his interest in astrology and astral influences described in the Hermetic texts brought him to the attention of a suspicious and sceptical Church. For Ficino, texts were the door to ancient knowledge and the key to recreating the Golden Age of classical antiquity. To his successors, however, texts were instructions—recipes for intervening in the real world.

The Neoplatonic seeds sown by Ficino at Cosimo's Academy in the 1450s were to flower in most of the courts of Europe but, above all, at Rudolf II's court in Prague. It was in Prague under Rudolf (1552–1612) that a spectacular scientific and cultural explosion occurred. From the 1580s, he immersed himself in the creation of the ideal Renaissance court, with vast collections, a Kunstkammer, gardens and a zoo. Rudolf was the model of the enlightened Renaissance monarch, gathering around him the brightest lights of his generation regardless of their religion:

GIUSEPPE ARCIMBOLDO, *Portrait of Emperor Rudolf II as Vertumnus*, *c.* 1590, TEMPERA ON WOOD, 70 X 58 CM. SKOKLOSTER, SKOKLOSTER CASTLE, 11615.

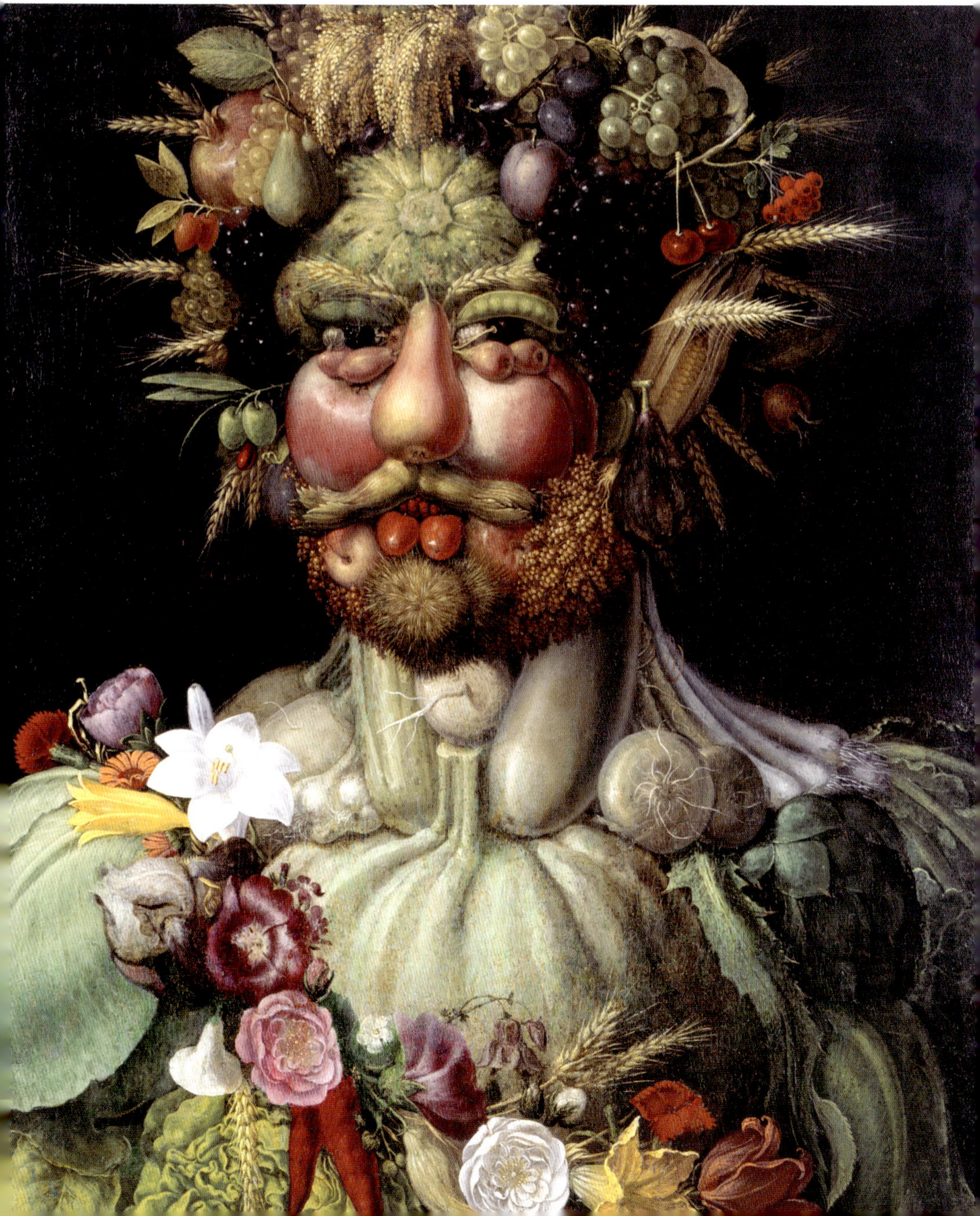

Spranger, De Vries, Hoefnagel, Brahe, Kepler, Brueghel, Habermel, Bürgi, Drebbel. In the years that followed, Rudolf, like Shakespeare's Prospero, gradually withdrew from public life, and "to [his] state grew stranger, being transported / And rapt in secret studies" (*The Tempest* 1. 2. 76–7). Like Prospero, he ended by seeing his kingdom in the hands of his brother, left only with his Kunstkammer, "dukedom large enough". His gradual withdrawal has left us a portrait of Rudolf II as an impotent recluse, obsessed with clockwork and crucibles, excluded from the exercise of power. However, Rudolf was seen by his contemporaries as the ruler of vast territories for over thirty-five years, victorious over the Turks, presiding over the most illustrious court of Europe.

Although Rudolf II died in 1612, after having abdicated in favour of his brother Matthias, the world was still largely a Habsburg world in 1620. European politics were dominated by the Spanish Habsburgs, who, along with their Austrian cousins, were on the thrones of Europe's most powerful countries. Philip II was on the Spanish throne, Ferdinand had routed the Protestant forces at the Battle of White Mountain near Prague, Rudolf II's brother Albrecht and his bride Isabella, Philip's daughter, ruled the Spanish possessions in the Low Countries. The Habsburgs dominated the European political arena, aggressively promoting the aims of the Counter-Reformation and trying to re-conquer Europe for the Roman Catholic faith. At the beginning of the Thirty Years' War, the Catholic Habsburg forces were in the ascendant. The Habsburgs were also powerful cultural models and Rudolf's court at Prague was emulated in Madrid, Vienna, Innsbruck and Antwerp.

This is the world we enter as we view the *Linder Gallery*, a world looking Janus-like both forward to Boyle's 'chymistry' and Newton's physics, and backward to Nostradamus' astrology and Sendivogius' alchemy. The 1620s was the world of Rubens, Brueghel, Van Dyck and Galileo, but it was also only recently the world of Shakespeare and Tycho Brahe. It was a world that stood at the threshold of the Thirty Years' War. It was a world alive with experiments and exploration, but also a world that remembered the Wars of Religion and the St Bartholomew's Day Massacre, the Spanish Armada and the assassination of Henry IV. This painting is a document that holds the clue to understanding the political, intellectual, artistic and scientific ferment of the first half of the seventeenth century.

The *Linder Gallery* was painted sometime in the 1620s and is one of a relatively modest number of similar paintings showing the interiors of private collectors' homes. This type of painting emerged in a single city—Antwerp—during the second decade of the seventeenth century. Images had a tormented role in the history of the Netherlands. During the iconoclasm of 1566, Calvinists stormed the churches of Antwerp desecrating and destroying paintings and sculptures representing saints and religious figures and beginning the Dutch Revolt, with the Protestant Northern Netherlands struggling for independence from the control of the Spanish Habsburgs. This prompted King Philip II of Spain to send the notoriously cruel Duke of Alba, nicknamed the Iron Duke, to quell the movement. In the ensuing bloodshed many thousands died and over a thousand Calvinists were executed. In 1585, after a second 'quiet' iconoclasm, Antwerp finally fell to Spain and Catholicism was triumphantly restored, along with an ambitious civic programme of art and architecture to rebuild what the iconoclasts had destroyed.

The Society of Jesus, founded by Ignatius of Loyola in 1534 and officially recognized as an order in 1540, rapidly established a power base in Antwerp second only to Rome, and the Plantin–Moretus press emerged as the publishing epicentre of the Counter-Reformation. Protestants living in Antwerp, once a city known for its religious tolerance, were provided with four years to either reconcile with Catholicism or leave. Many left and the population dropped from 105,000 in 1568 to only 42,000 in 1589. In the same year the United Provinces of the Northern Netherlands managed to cut off Antwerp's primary access to trade by blockading the river Scheldt, marking the decline of what had once been Europe's primary trading centre and permitting the inexorable rise of Amsterdam. This is how Antwerp was described in 1627 by its most famous artist, Peter Paul Rubens:

> This city ... languishes like a consumptive body, declining little by little. Every day sees a decrease in the number of inhabitants, for these unhappy people have no means of supporting themselves either by industrial skill or by trade. One must hope for some remedy for these ills caused by our own imprudence.

In 1598, Philip II sent his favourite daughter Isabella and her husband Archduke Albrecht of Austria, to govern the Netherlands. The Archdukes, as they were known, set up their court in Brussels. After years of war and significant victories by the Protestant forces, a ceasefire was finally achieved in 1609, beginning the Twelve Years' Truce. This period of peace signalled an extraordinary new flourishing of the arts in Antwerp. When the Archdukes invited Rubens back from his travels in Italy to become their court painter, he accepted on condition that he could set up his workshop in Antwerp rather than the archducal residence of Brussels. From 1610, Rubens' home rapidly became an artistic and cultural Mecca frequented by artists including Jan Brueghel the Elder, Anthony van Dyck, Frans Francken II, Hendrick van Balen and many others. The new Jesuit church and other religious orders brought into Antwerp by the Archdukes provided Rubens' studio with a host of painting commissions, while the Plantin–Moretus press required the most skilled engravers to create the visual propaganda of the Counter-Reformation and the illustrated books published by Jesuit scholars.

From the late sixteenth to the early seventeenth century, the creation of private art collections became increasingly popular among wealthy Antwerp merchants. The private collection became a civilized place to accumulate, display and discuss paintings and precious objects, away from the devotional context of churches. The most important Antwerp painters' guild—the Guild of St Luke—admitted 'lovers' of art—*liefhebbers* or connoisseurs—as members from the first decade of the seventeenth century, along with practising artists. The ability to discern the style of different artists and to unravel complex allegories in a game of polite conversation became defining social skills for wealthy Flemish burghers.

Accompanying the rise of a new kind of intellectual private collector was an elevation of the social status of the artist. Whereas painters were often considered as socially inferior craftsmen in the sixteenth century, Rubens and the many Flemish artists that orbited his studio took pains to argue the nobility of painting, both in their dress and social behaviour and in elaborate allegorical works that emphasized the status of painting as one of the liberal arts. Rubens and his most talented apprentice Van Dyck attracted particular attention for their aristocratic attire. Rubens' own house contained an extensive collection of artworks, antiquities and natural curiosities, further blurring the lines between artist and collector.

It was in this environment—in Antwerp during the peaceful interlude of the Twelve Year's Truce and the decadent *ennui* of a city based increasingly on the trade in luxury goods—that a completely new kind of painting was born: the gallery interior. The earliest examples, dating from the early years of the second decade of the seventeenth century, are associated with the studio of Frans Francken II. Rather than representing faithful visual inventories of actual collections, they depict idealized microcosms in palatial spaces quite unlike the low-ceilinged private buildings of Antwerp—spaces filled with paintings, antique sculpture, medals and natural wonders, including exotic shells and flowers. These encyclopaedic collections are often populated by allegorical figures in classical garb. Some gallery interiors include connoisseurs and scholars discussing the merits of different pictures or inspecting globes and maps. Living dogs, monkeys and parrots are frequent visitors to the gallery space, and the Archdukes make numerous cameo appearances in the gallery along with celebrated Flemish artists and Antwerp art lovers as well as reanimated humanist scholars such as Justus Lipsius and Erasmus, often with architectural settings alluding to Rubens' gallery or implausible vistas of the archducal residences. The most famous examples of the gallery interior genre are the paintings comprising the *Five Senses* series, created by Rubens and Jan Brueghel the Elder in 1617 for the Archdukes and now in the Prado.

These works were almost always collaborations and they defy the notion of the heroic individual artist, being more like ambitious cinematic or theatrical productions with a full list of credits (landscape by Bril, flowers by Brueghel, figures by Van Balen, concept by Rubens, etc). A key part of the appreciation of these works lay in the unravelling of the hands of the different artists involved, in the identification of their contents and in deciphering their playful allegories. Their collaborative nature contributed to their value. Many Antwerp artists were also members of the chamber of rhetoric 'De Violieren' and delighted in the creation of emblems and puzzles. It was only later in the seventeenth century, particularly in the works of David Teniers the Younger, that the allegorical figures and natural curiosities disappeared from the gallery space, and these paintings became faithful portraits of existing collections rather than visual encyclopaedias designed to trigger a particular kind of learned conversation.

The *Linder Gallery* clearly belongs to this evolving genre of the Antwerp gallery interior. While it is almost certain that this work was painted by an artist or artists from Antwerp, likely connected to the workshop of Jan Brueghel the Elder, the work was commissioned by a wealthy German merchant living in Milan, Peter

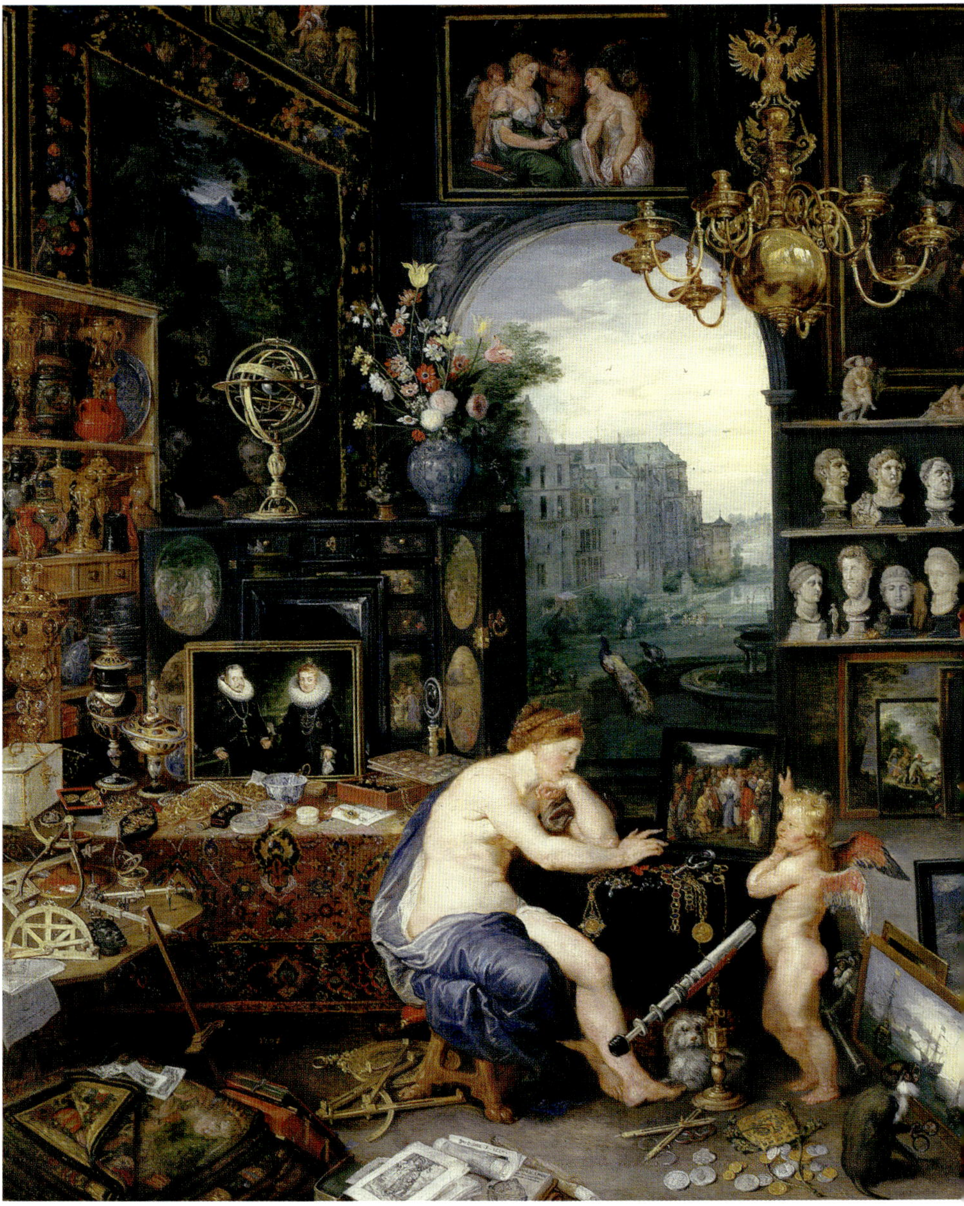

PETER PAUL RUBENS (SIEGEN 1577–ANTWERP 1640), JAN BRUEGHEL THE ELDER (BRUSSELS 1568–ANTWERP 1625), *Allegory of Sight*, 1617, 65 x 109 CM. MADRID, PRADO, P01394.

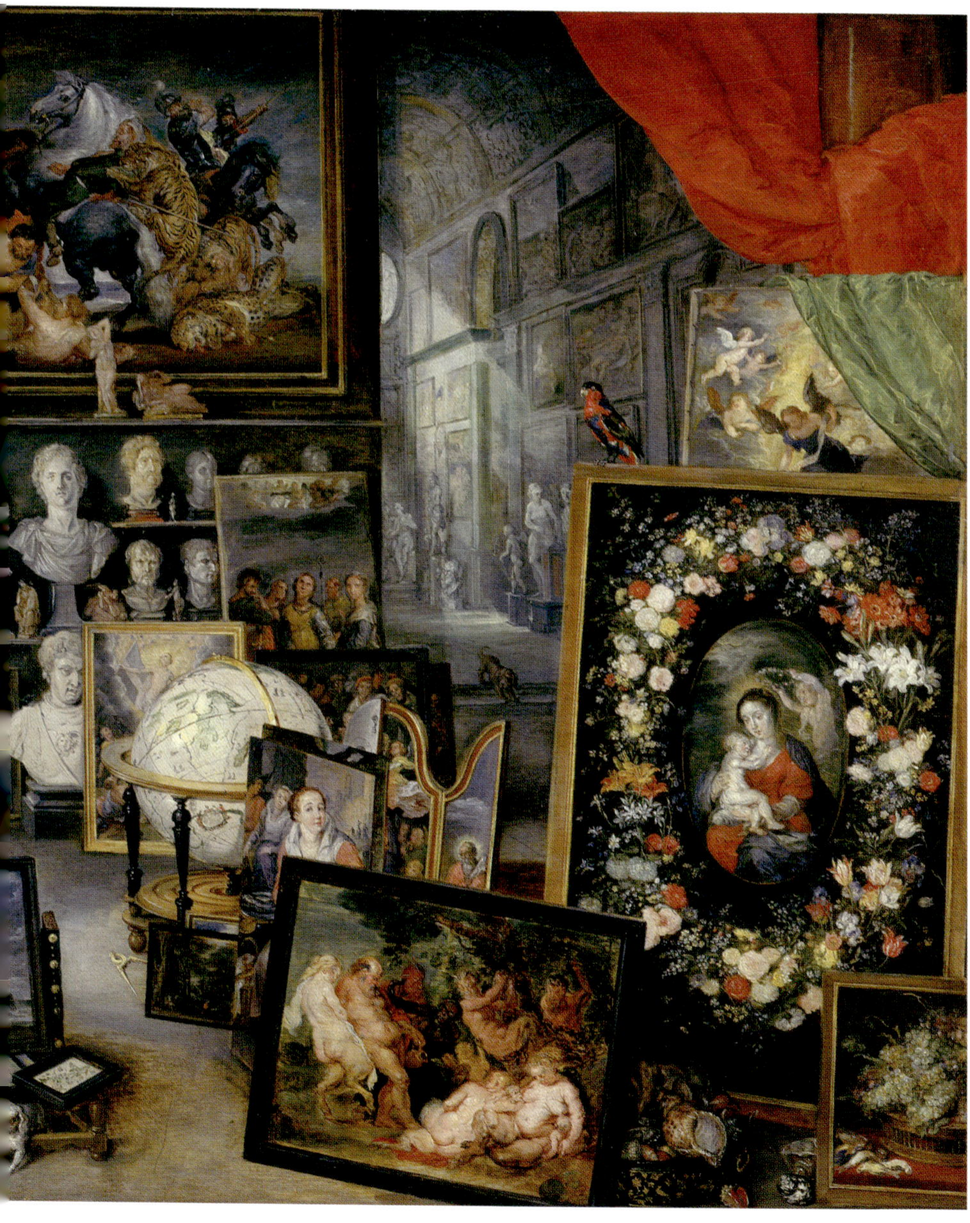

Linder, probably in the early 1620s. We know that Linder's friend and mathematics tutor, the Urbino mathematician and architect Muzio Oddi, whose portrait medal appears almost illegible in the centre of the painting, played a key role in informing the composition of the picture. The archbishop of Milan, Cardinal Federico Borromeo, was a major collector of Flemish paintings and was particularly fond of the works of Jan Brueghel the Elder. The movement of Flemish artists and works between Antwerp and Milan was facilitated by the frequent journeys back and forth of Borromeo's agent Ercole Bianchi, another pupil of Oddi.

The *Linder Gallery* was most probably conceived in Milan in the early 1620s, on the basis of a preparatory drawing now in the Royal Collection in Windsor Castle. The painting depicts the interior of a gallery in perspective, with a view over an Italianate garden with a fountain and an obelisk. The walls are adorned with the paintings of Flemish and Italian artists, and the room also contains antique sculptures and more recent works. Three tables in the foreground bear astronomical and mathematical instruments, drawing implements and diagrams. Two allegorical figures dominate the foreground—an old man and, reclining in his lap, a young woman holding paintbrushes, palette and a maulstick, a stick with a padded end used by painters to support the hand holding the brush, as well as a sculptor's mallet. The woman wears a laurel wreath and has a sun pendant around her neck.

The concentrated intensity of the two foreground figures is remarkable, particularly when considered in contrast to the usual conversing connoisseurs or voluptuous nudes of Antwerp gallery interiors of the period. But who are they? Allegorical personifications? Or portraits of real individuals? Is the woman alive or dead? What is her relationship to the man? These are questions to explore in the conversation, but there is still much uncertainty around them. She seems to represent *Pictura*—painting and the arts—and he *Disegno*—drawing, design, measurement and intellectual apprehension of the world, frequently described as the father of painting. It is also possible that the male figure, as an embodiment of mathematical skill, is intended as a likeness of German astronomer Johannes Kepler (1571–1630). If the Rubens–Brueghel *Five Senses* series celebrates our ability to delight the senses through the bounty of nature and the fruits of art, the *Linder Gallery*, commenced shortly after that ambitious production, could be considered as a 'sixth sense'—the intellectual understanding of the cosmos through measurement and mathematics.

Hans von Aachen (Cologne 1552 –Prague 1615), *Portrait of Kepler* (?), *c.* 1612, oil con canvas, 51.5 x 38.5 cm. Rychnov nad Kněžnou, Zámecká Obrazárna větve Kolowratů, Kolowrat Collection.

Following page
Flemish school, *A Picture Gallery*, *c.* 1620–30, pen and ink with wash over graphite, 397 x 606 mm. Windsor Castle, RL 12983.

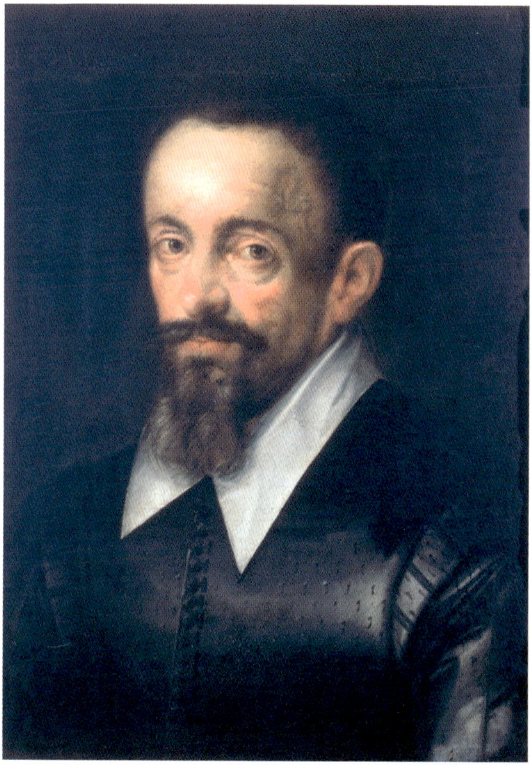

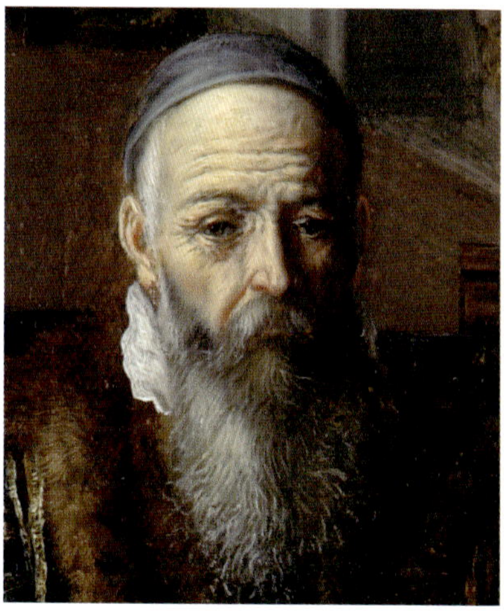

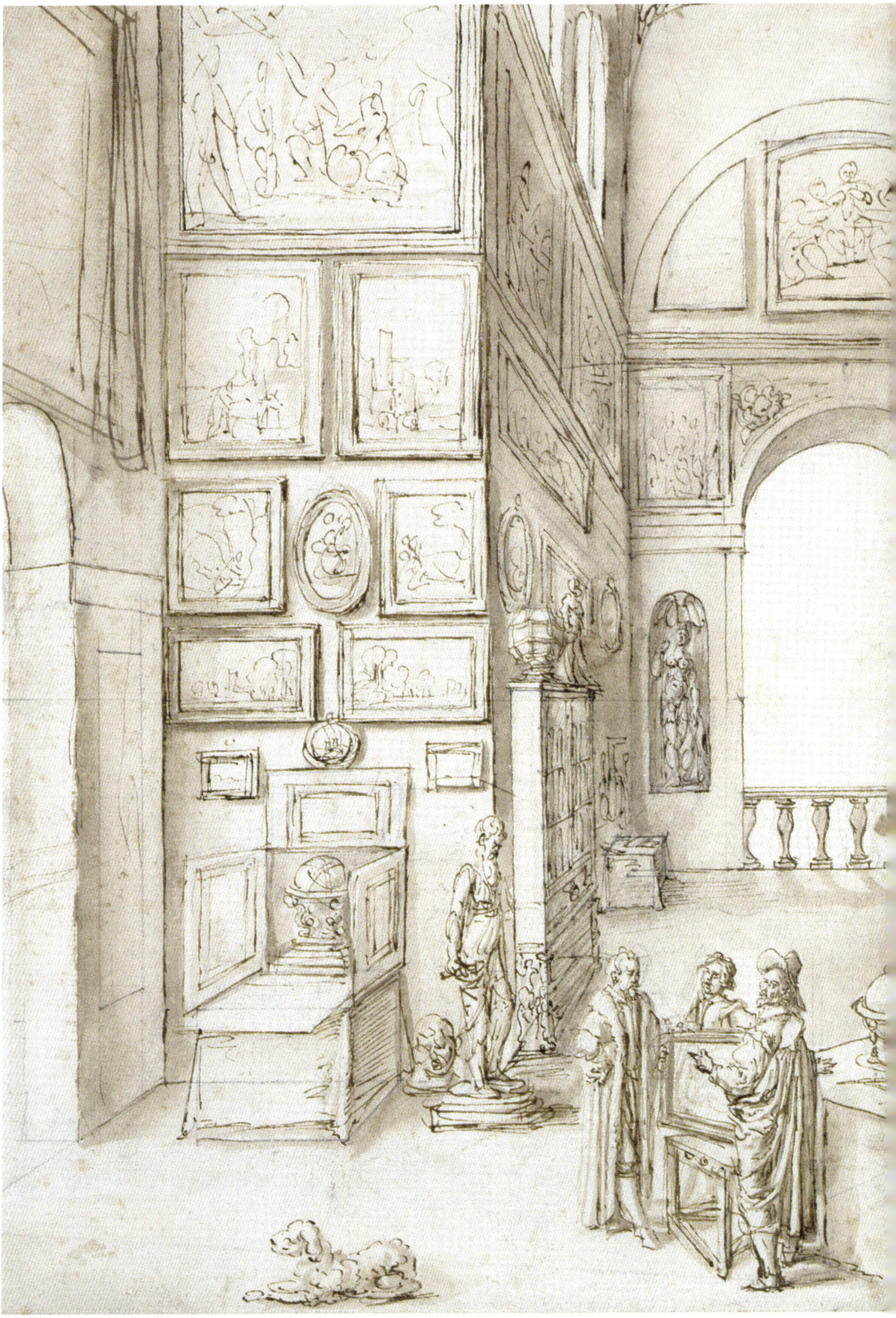

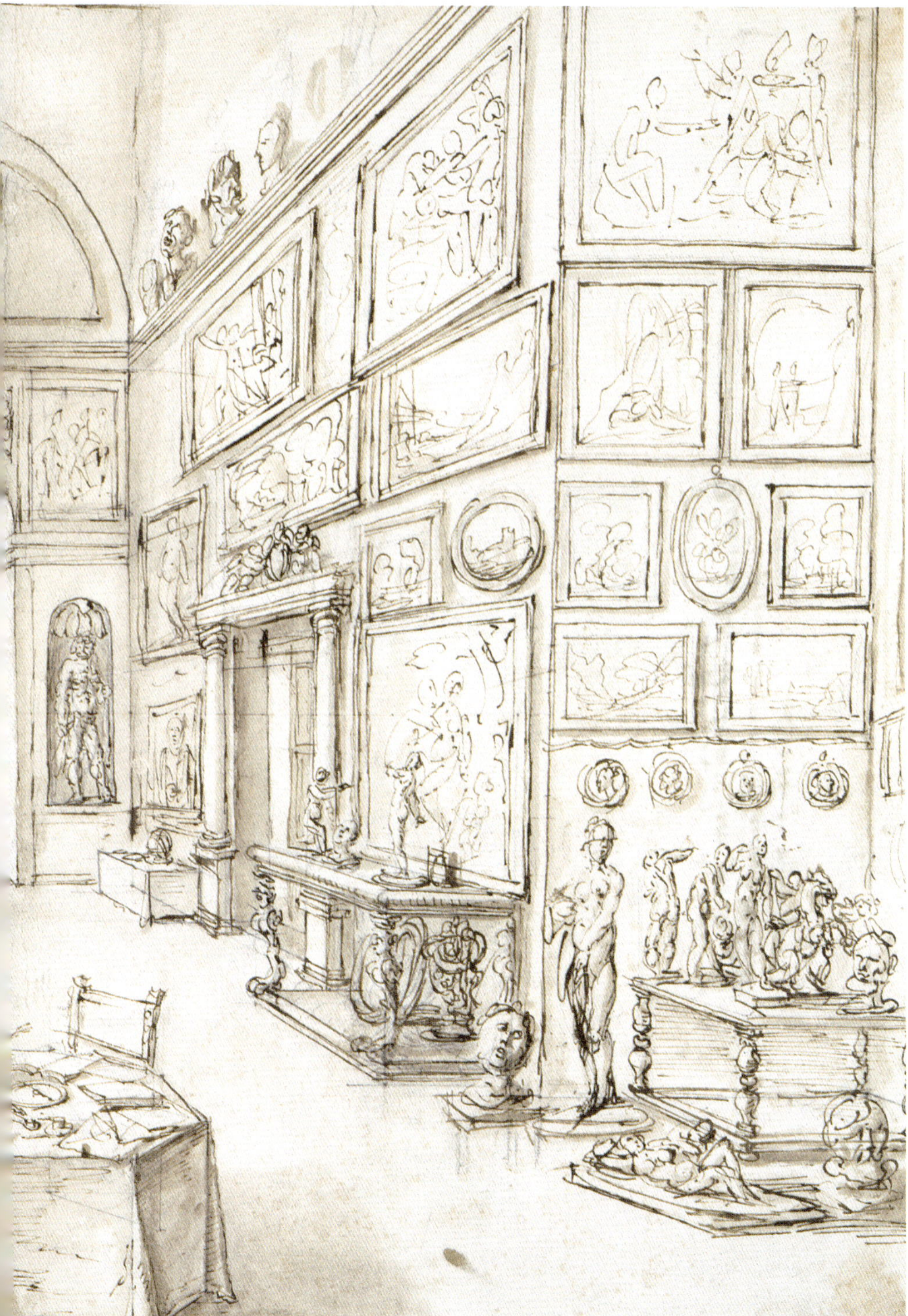

The *Linder Gallery* stands apart from its genre in a number of ways. In addition to the extraordinary detail of the execution, the floor is empty of the dogs, monkeys and flowers typical of Flemish gallery interiors. But perhaps most intriguing are the references to the cosmological debates of the seventeenth century. It was not unusual for gallery interiors to contain astronomical instruments—Rubens' and Brueghel's *Allegory of Sight* contains a telescope, an armillary sphere and various other instruments, and globes and compasses are common features in the interiors, particularly those associated with the Brueghel and Francken workshops. Mathematical skills were a key element in the training of Catholic nobles in the early seventeenth century, particularly under the guidance of the Jesuit colleges, and a rich assortment of astronomical and mathematical instruments signed by the most famous instrument-makers of the day was an obligatory element of any private gallery in Antwerp or Milan. The ability to identify the scientific instruments in a painting could be as important a status symbol as the ability to identify the artworks, and indeed there was considerable overlap between the practitioners and workshops involved in painting, engraving and instrument-making.

However, the *Linder Gallery* draws us deeper into the cosmological discussions of the seventeenth century than any other painting of this kind. In addition to the extraordinarily precise rendering of existing instruments such as the large astrolabe by Gualterus Arsenius and the celestial globe by Hondius, the octagonal table includes a remarkable reference to the dispute over the structure of the universe, on a piece of paper bearing the three competing cosmic systems: the Earth-centred universe of Ptolemy, the Sun-centred universe of Copernicus and the compromise system of the Danish nobleman and astronomer Tycho Brahe (1546–1601), in which the inner planets orbited the Sun but the Sun and the other planets orbited the Earth. Beneath the diagrams of the cosmos and written in minute, precise lettering is the cryptic Latin inscription ALY ET ALIA VIDENT, meaning 'Different people see it differently' or 'Others see it yet otherwise'. Why is this diagram so central to the painting, almost appearing to come off the table as it is pushed towards the viewer? Another clue is presented by the three books on the octagonal table to the right of the globe. These can be identified as the *Harmony of the World* (1619) by Johannes Kepler, the *Rudolphine Tables* (1627) also by Kepler, and the *Description of the Admirable Table of Logarithms* (1614) by Scottish mathematician John Napier (1550–1617).

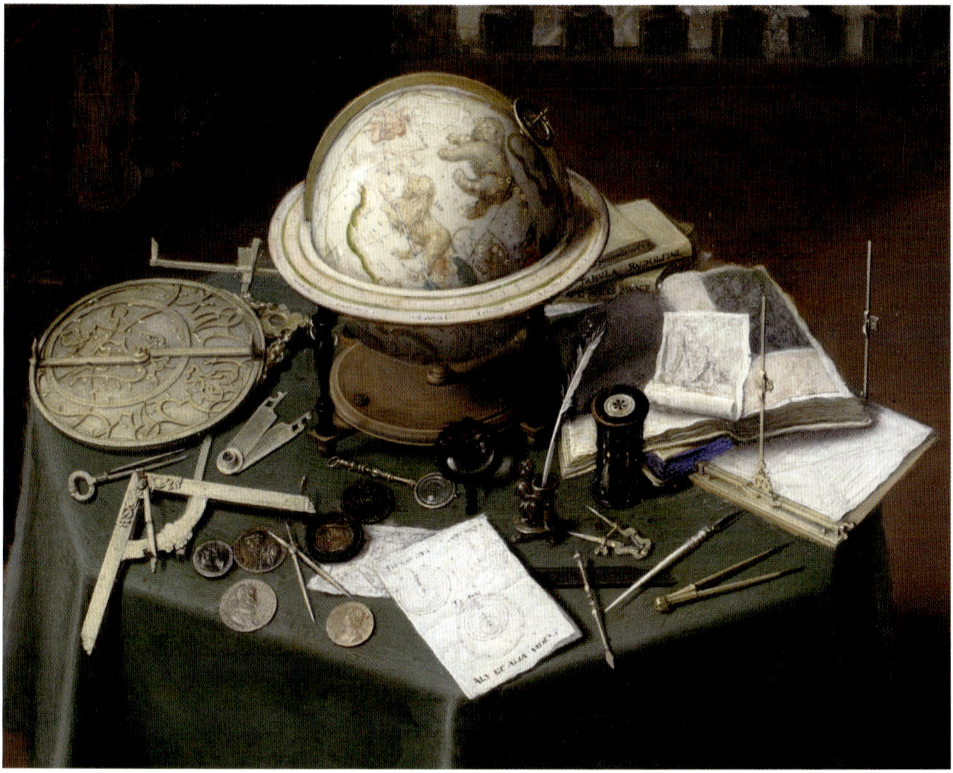

To understand these clues we need to explore some of the defining moments in the history of astronomy in the period leading up to the *Linder Gallery*. The year 1543 saw the publication of Nicolaus Copernicus' *On the Revolutions of the Heavenly Spheres*, proposing a Sun-centred universe. Copernicus' mathematical arguments emphasized that while the idea of the Earth's motion might seem highly counter-intuitive based on our everyday experience, the Sun-centred solution was more elegant than the Earth-centred system of Claudius Ptolemy. An anonymous preface, inserted into Copernicus' work while he was on his deathbed, suggested that the heliocentric or Sun-centered hypothesis should not be taken seriously as a claim about the nature of the universe, but merely viewed as a convenient fiction—a useful calculating device.

In spite of its controversial premise, Copernicus' work had a limited readership due to its technical content and went largely unnoticed beyond a small group of astronomers for several decades. Copernicus did not present significant new em-

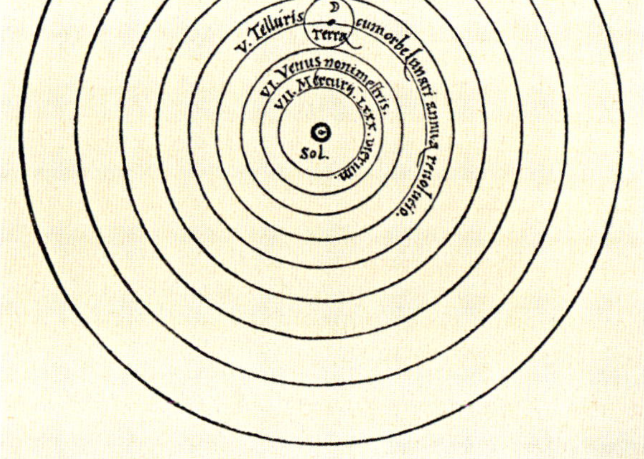

THE COPERNICAN SYSTEM, FROM NICOLAUS COPERNICUS, *De revolutionibus orbium coelestium*, KRAKOW, JAGIELLONIAN LIBRARY, MS 10000, FOL. 9V.

pirical data, but merely a hypothesis based on mathematical elegance. The prevailing Aristotelian view of the world taught in sixteenth-century European universities held that the Earth was at the centre of the universe and that the planets were carried around it on solid crystalline spheres. The outermost sphere was occupied by the so-called 'fixed stars'. Each of the four elements strove to occupy its natural place in the universe, with earth occupying the lowest point, followed by water, air and fire in concentric spheres around the centre of the Earth. These 'sublunar' elements were capable of change and mutation, but once you arrived at the sphere of the Moon you entered the realm of the 'quintessence' or ether, where no change was possible and the Moon, planets and fixed stars rotated around the Earth in eternal circular perfection. After the Council of Trent ended in the 1560s, Aristotelian philosophy and the theology of Thomas Aquinas were enshrined as the foundations of reformed Catholic doctrine.

While it was beautifully simple, Aristotle's theory had a number of problems.

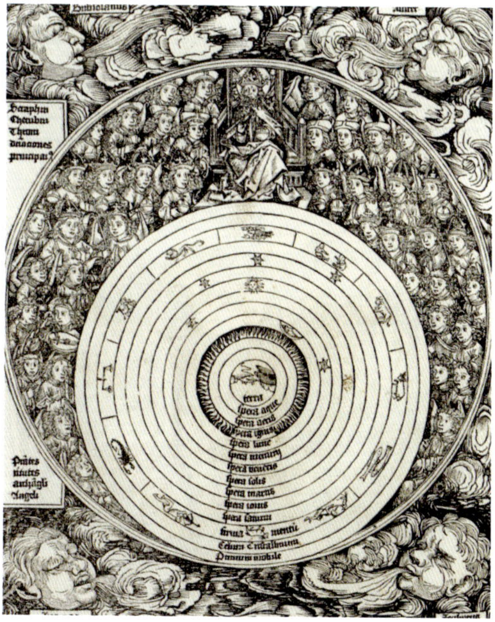

The Ptolemaic Universe, FROM HARTMANN SCHEDEL, *Das Buch der Croniken und Geschichten*, NUREMBERG, ANTON KOBERGER, 1493. FLORENCE, BIBLIOTECA RICCARDIANA, ED. RARE 233, FOL. Vv.

Firstly, the planets did not in fact appear to rotate in regular circles around the centre of the Earth. Instead they stopped and started and occasionally even appeared to move backwards. To deal with these problems the Alexandrian astronomer Claudius Ptolemy (*c.* AD 100–*c.* 170) used a number of awkward devices including the epicycle, a wheel attached to the rotating crystal sphere on which the planet rotated, the eccentric, in which the Earth's centre was displaced from the centre of the planet's orbit, and the equant, an imaginary point from which the planet appeared to move at a constant speed. These devices were introduced as 'convenient fictions' rather than true physical structures. While Copernicus managed to do away with Ptolemy's equant, his theory still contained epicycles and eccentrics, and was still based on the idea that the planets were carried around the Sun on solid crystalline spheres.

In 1572 Tycho Brahe observed a new star in the sky, a supernova. This appeared to challenge the fundamental Aristotelian belief that no change was possible in the heavens, and inspired him to engage on an extraordinary programme of observational astronomy. At vast expense, Tycho created Uraniborg, an observatory equipped with large, extremely precise astronomical instruments on the Danish island of Hven, embarking on a monumental programme of empirical data collection of planetary and star positions. After the death of his patron King Frederick II of Denmark and the decline of his influence under his successor Christian, Tycho moved to Prague in 1599 to the court of Emperor Rudolf II to continue his observations of the stars and planets. Tycho proposed a system of planetary orbits that was something of a compromise between the Ptolemaic system and the counter-intuitive Copernican system, with Mercury and Venus orbiting the Sun but the Sun itself, along with the outer planets and fixed stars, orbiting the Earth. His system abandoned the idea that the planets rotated on crystalline spheres, instead having them moving "like birds in the air or fish in the water", under the influence of some as yet unknown force.

Tycho's talented and mercurial assistant in compiling his astronomical data at Rudolf's court was Johannes Kepler. In 1595 Kepler, a convinced Copernican, was considering the sizes of the orbits of the planets and had a sudden epiphany—it seemed that the orbits of the seven planets could fit perfectly inside the five perfect or Platonic solids. Kepler published the results and the extraordinary visual image of his Mannerist universe, which he also proposed as the design for a novel drinks dis-

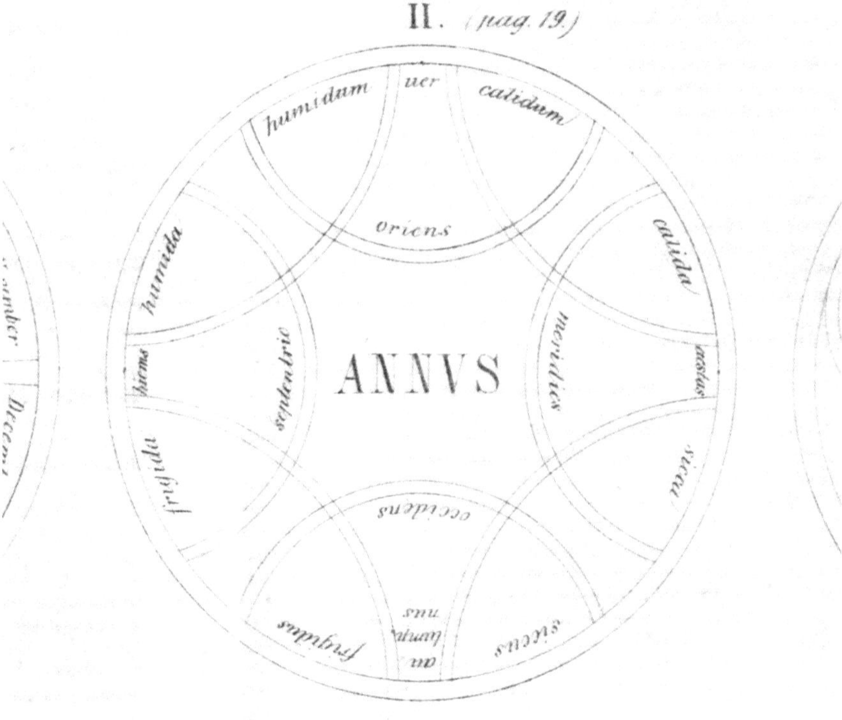

TETRAGRAM, FROM ISIDORE OF SEVILLE, *De responsione mundi et astrorum ordinatione* (*De natura rerum*), AUGSBURG, GÜNTHER ZAINER, 1472.

penser, in his *Mysterium cosmographicum*, the 'Cosmographic Secret', which he sent to Tycho. While Tycho did not take Kepler's revelation about the orbits of the planets particularly seriously, he was impressed by his mathematical ability and set him to work on the preparation of his astronomical tables. It was Tycho's data for the planet Mars that presented Kepler with the most trouble. After wrestling with it for some time, Kepler hit on the idea that the planetary orbits might be described by ellipses rather than circles, and eventually developed his three laws of planetary motion, proposing a new cosmic physics where the planets were moved by a magnetic influence from the Sun, sweeping out equal areas in equal times.

Kepler was fascinated by the ancient Pythagorean quest for the 'music of the spheres', the musical notes supposedly produced by the different planets as they meandered through the heavens. In his most profound exploration of cosmic music, *The Harmony of the World* published in 1619 and present on the central table of the *Linder Gallery*, he remarked that "The Earth sings *Mi, Fa, Mi*: you may infer even from the syllables that in this our home *Mi*sery and *FaMi*ne hold sway", referring to the wretched state of the German lands in the early seventeenth century. Kepler's cosmic polyphony was based on the fact that the ratios of the maximum and minimum speeds of planets on neighbouring orbits very closely approximated musical intervals, which Kepler investigated with mathematical precision. As well as advancing Kepler's Third Law of planetary motion, the *Harmony of the World* also dealt extensively with astrology. Like Tycho Brahe before him and most other astronomers of the time, Kepler was partly dependent on astrology for his income and cast many horoscopes for his patrons, while seeking to develop a more solid foundation for astrology in the laws of physics. Kepler's *Rudolphine Tables*, publishing Tycho's accurate star positions and tables for the planets, were anxiously anticipated by astrologers throughout Europe. In the early decades of the seventeenth century the European witch-craze was at its height and Kepler was delayed in completing the tables when he had defend his own mother, Katherina, who was accused of being a witch. The *Rudolphine Tables*, dedicated to Tycho's patron the late Emperor Rudolf II, were finally published only in 1627, shortly followed by a supplement written by Kepler to facilitate astrological use of the tables.

The elaborate frontispiece of the *Rudolphine Tables* shows the Temple of Urania, the muse of astronomy. In the temple, accompanied by Copernicus, Ptolemy and

Great Mural Quadrant, FROM TYCHO BRAHE, *Astronomiae instauratae mechanica*, WANDESBURG, PHILIP DE OHR, 1598. FLORENCE, BIBLIOTECA NAZIONALE CENTRALE DI FIRENZE, PALAT. 8.9.7.30, FOL. VIIIV.

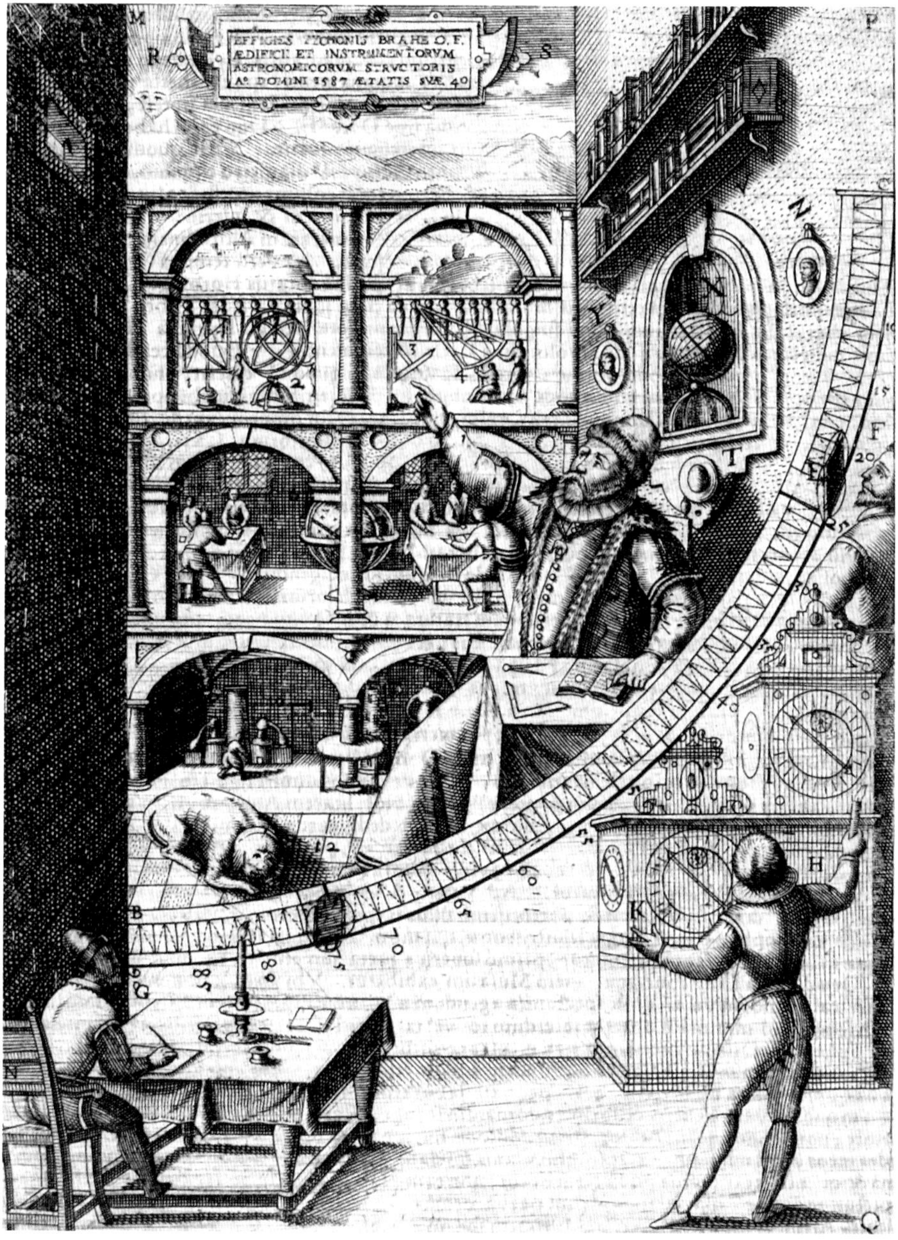

ancient astronomers, Tycho can be seen pointing up towards the ceiling where his system of the universe is depicted and uttering the words *Quid si sic?* ('What if it is thus?'). In the base of the temple, beside Tycho's island of Hven, we see a portrait of the ageing Kepler in a cramped space, next to an architectural model of the roof of the temple showing him as the architect of the new astronomy, performing calculations and waiting for a few coins to drop from the Habsburg double-headed eagle at the top of the temple. Kepler's complex calculations of planetary tables were made far easier thanks to the discovery of logarithms by Scottish mathematician John Napier. Kepler emphasized his enormous debt to Napier in the *Rudolphine Tables*, and Napier's book, recently identified by Paolo Galluzzi, appears conveniently on top of Kepler's works in the *Linder Gallery*.

Another recipient of a copy of Kepler's revelations in *Mysterium cosmographicum* was Galileo Galilei (1564–1642), who also confided to Kepler in 1597 that he was a Copernican. When Galileo adapted the Dutch invention of the telescope and

Georg Celer, *Temple of Urania*, frontispiece of Johannes Kepler's *Tabulae Rudolphinae*, Ulm, Jonas Saur, 1627. Florence, Biblioteca Nazionale Centrale di Firenze, Magl. 5.1.118.

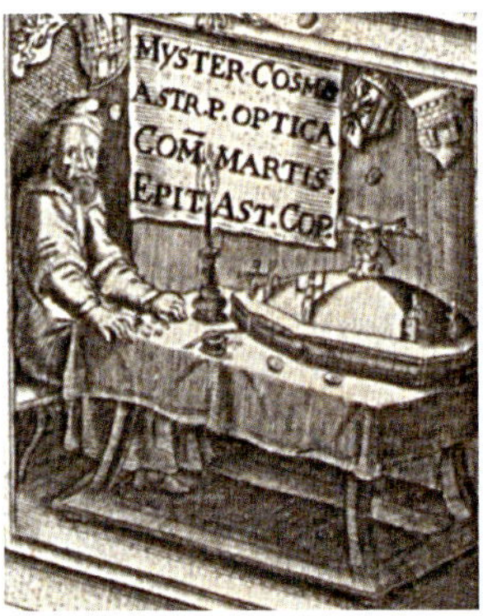

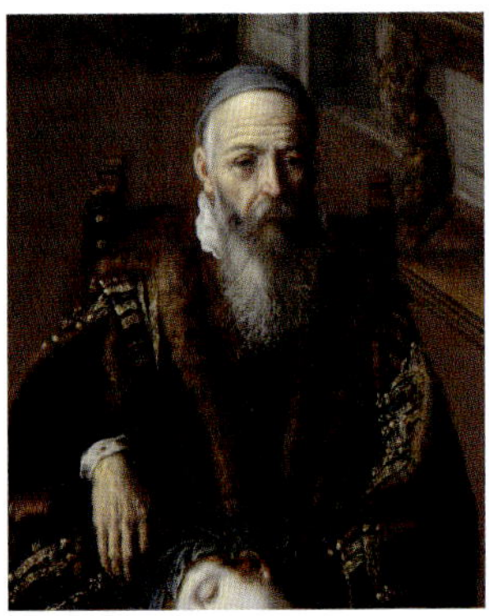

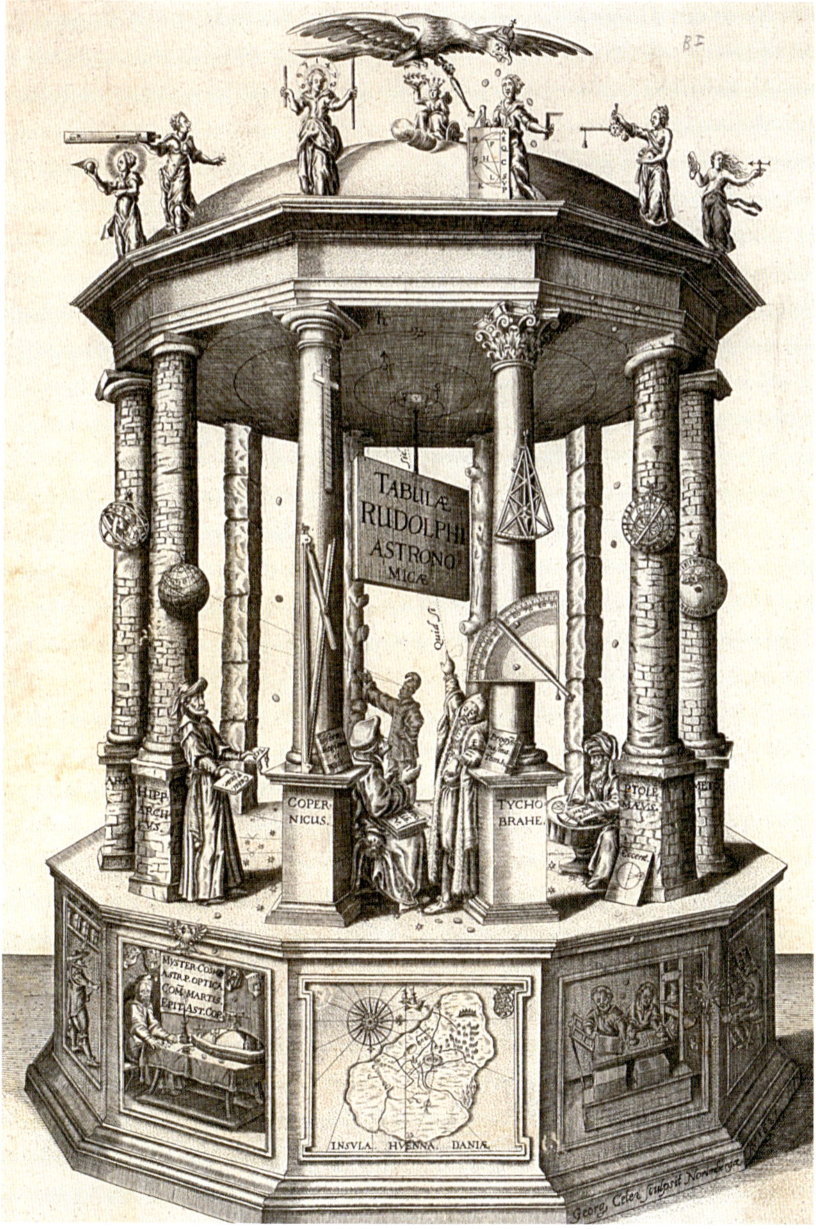
TABULÆ RUDOLPHI ASTRONOMICÆ
COPERNICUS.
TYCHO BRAHE.
HIPPARCHUS.
PTOLEMAUS.
MYSTER.COSM.
ASTR.P.OPTICA
COM.MARTIS
EPIT.AST.COP.
INSULA HUENNA DANIÆ.
Georg. Celer Sculpsit Norimbergæ

began using it for astronomical observations in 1609, he observed the rough and mountainous surface of the Moon, providing additional evidence against Aristotle's theory that celestial bodies were perfectly spherical, and four tiny dots of light regularly orbiting Jupiter, which demonstrated conclusively that not everything in the solar system went around the Earth. He also saw that Venus had phases like the Moon, showing that it orbited the Sun and not the Earth. All of these observations would have been compatible with the Tychonic system, but Galileo had additional arguments for the Earth's motion—a shaky one based on the behaviour of the tides, which Galileo compared to water sloshing around in a basin as the Earth rotated, and a more convincing one based on his observations of the movement of sunspots. Remarkably, Galileo studiously ignored both Tycho's system and Kepler's ellipses in his writings on the Copernican question. Ptolemy was probably an easier straw man to defeat than Tycho Brahe and Galileo appears to have been aesthetically attached to circular motions.

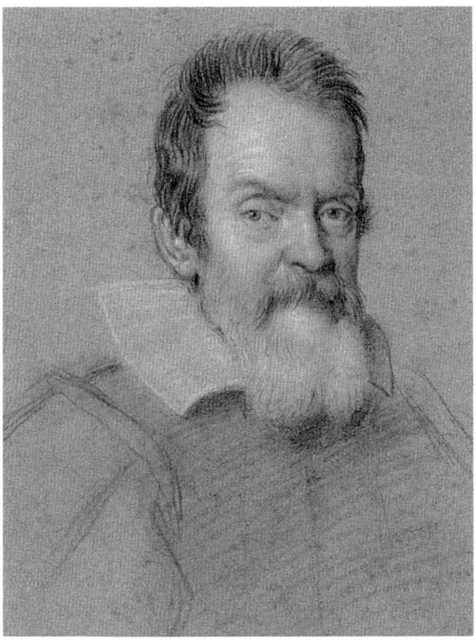

OTTAVIO LEONI, *Portrait of Galileo Galilei*, 1624, BLACK CHALK, RED CHALK, WHITE HEIGHTENING ON LIGHT BLUE PREPARED PAPER, 237 X 165 MM. FLORENCE, BIBLIOTECA MARUCELLIANA, VOL. H DISEGNI.

GALILEO GALILEI, *Drawings of the Moon*, 1609, AUTOGRAPH WATERCOLOUR. FLORENCE, BIBLIOTECA NAZIONALE CENTRALE DI FIRENZE, MS GAL. 48, FOL. 28R.

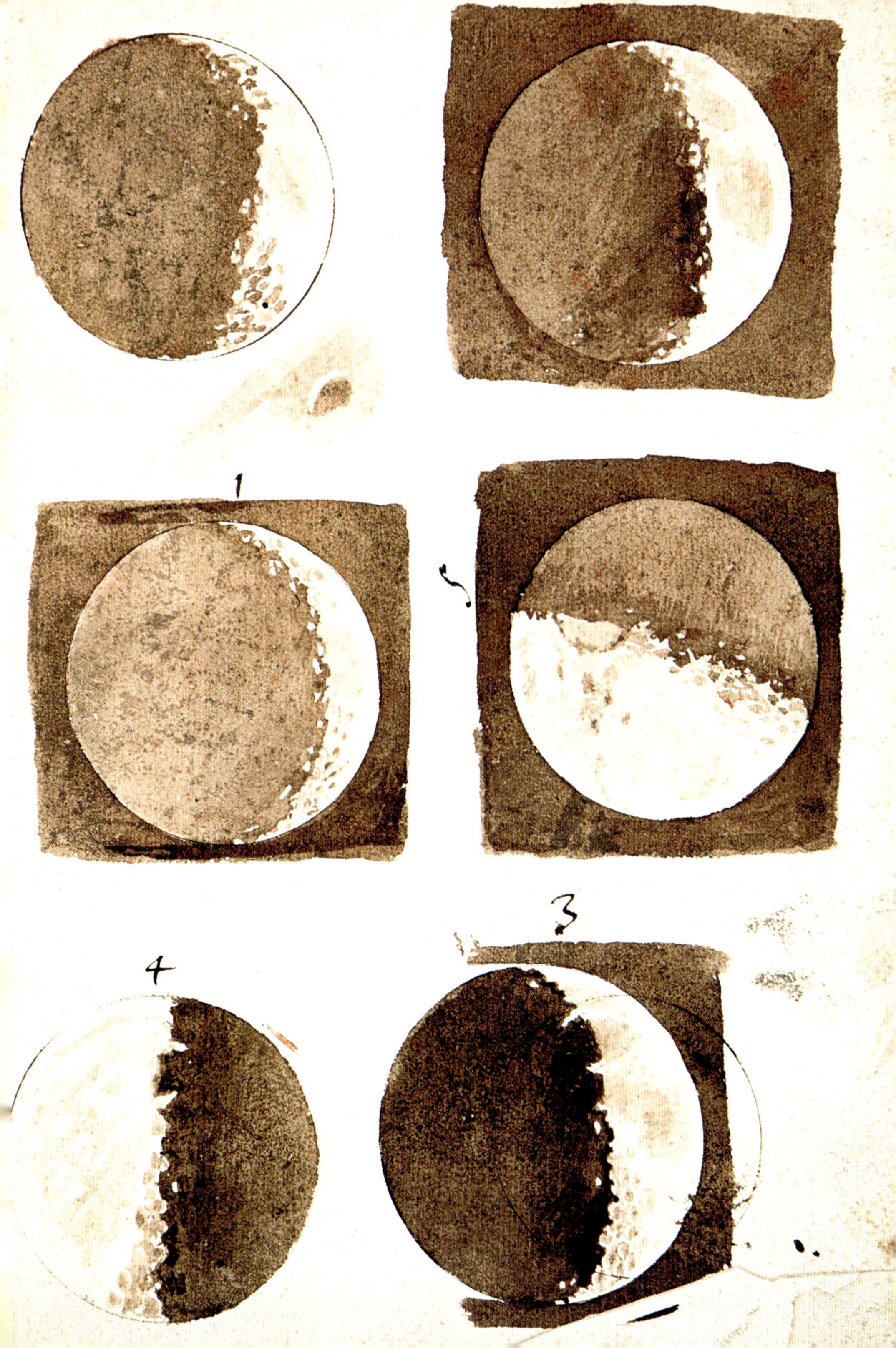
1
5
3
4

Galileo's arguments in favour of the Copernican theory and his suggestions that the passages from Holy Scripture that seemed to suggest that the Earth was stationary should be reinterpreted in the light of his findings were viewed as a challenge by the theological authorities. He was summoned to Rome by the Inquisition in 1616 and admonished not to teach or defend the Copernican theory. Copernicus' book was placed on the *Index of Prohibited Books* "until corrected", over seventy years after its original publication.

More than the details of Copernican theory itself, however, what was really subversive in the discussions around the solar system that were intensifying in the first decades of the seventeenth century was the fact that mere astronomers like Galileo and Kepler claimed to have true knowledge of the structure of the universe, usurping the rival claims of the Aristotelian philosophy so beloved of universities and even daring to encroach on the sacred domain of scriptural interpretation. Like artists, astronomers were involved in a battle for status in the early seventeenth century. Astronomy, geometry, arithmetic and music—the so-called *quadrivium*—were combined with the 'trivial' subjects of logic, rhetoric and grammar to make up the seven liberal arts taught in early modern schools: they were considered useful practical skills rather than certain knowledge. Bringing to bear empirical observations and experiments, mathematicians challenged traditional physics or 'natural philosophy' as taught through rational argument from everyday experiences. Galileo's insistence on holding the combined titles of 'mathematician and philosopher' to the Grand Duke of Tuscany when he moved to the Tuscan court was a provocative in-

Galileo Galilei, *Drawings of Sunspots*, c. 1612. Florence, Biblioteca Nazionale Centrale di Firenze, MS Gal. 57, fol. 69r.

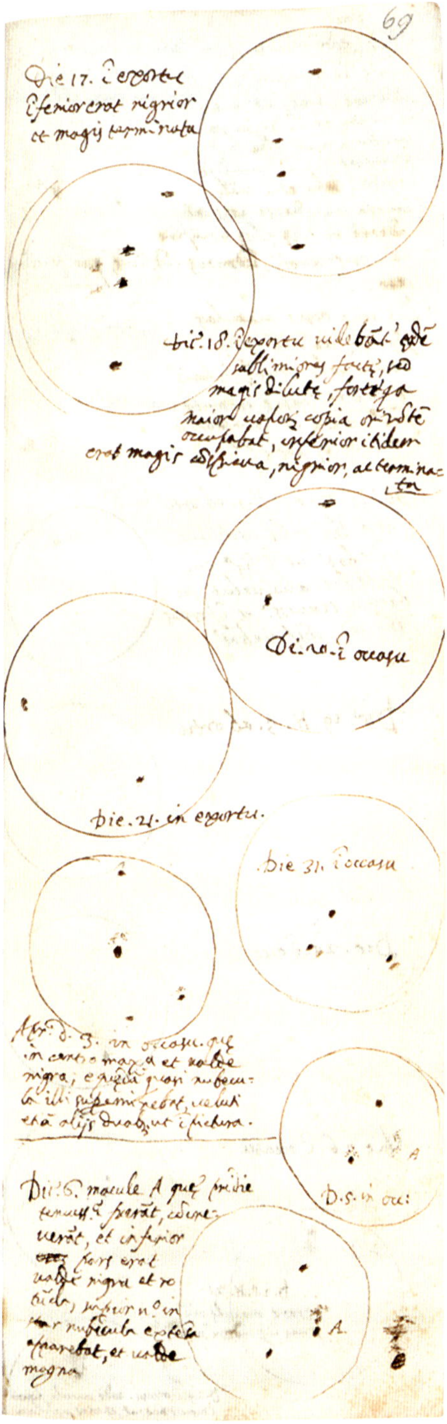
69
Die 17. i exortu
Die 21. in exortu
Die 31. i occasu
D. 5. in occ:

dication of the social climbing of Italian mathematicians. He made the claim most memorably in his 1623 book *The Assayer*, arguing that

> Philosophy is written in this grand book—the universe—which stands continuously open to our gaze. But the book cannot be understood unless one first learns to comprehend the language and interpret the characters in which it is written. It is written in the language of mathematics, and its characters are triangles, circles, and other geometrical figures...

It is hard for us with hindsight to appreciate just how radical it was at the time to believe that we could understand the world through mathematics. From black holes to genomics, from superstrings to microwaves, we have become accustomed to the concept of mathematics as a universal language of nature. Could the *Linder Gallery* be intended as a polemical presentation of precisely this view, that the world is amenable to understanding through numbers?

ALY ET ALIA VIDENT, the central 'slogan' of the *Linder Gallery*, is somewhat ambiguous, though. 'Others see it yet otherwise', the most accurate translation, would seem to imply that in addition to Ptolemy, Copernicus and Tycho, others propose different theories of the universe, a plausible nod to Kepler's ungainly elliptical orbits. However, 'Different people see it differently', also a possible translation, is altogether more non-committal during a period in which the cosmic systems were a

STEFANO DELLA BELLA, FRONTISPIECE OF GALILEO GALILEI'S *Dialogo sopra i due massimi sistemi del mondo, tolemaico, e copernicano*, FLORENCE, GIOVAN BATISTA LANDINI, 1632. FLORENCE, BIBLIOTECA NAZIONALE CENTRALE DI FIRENZE, B.R. 171.

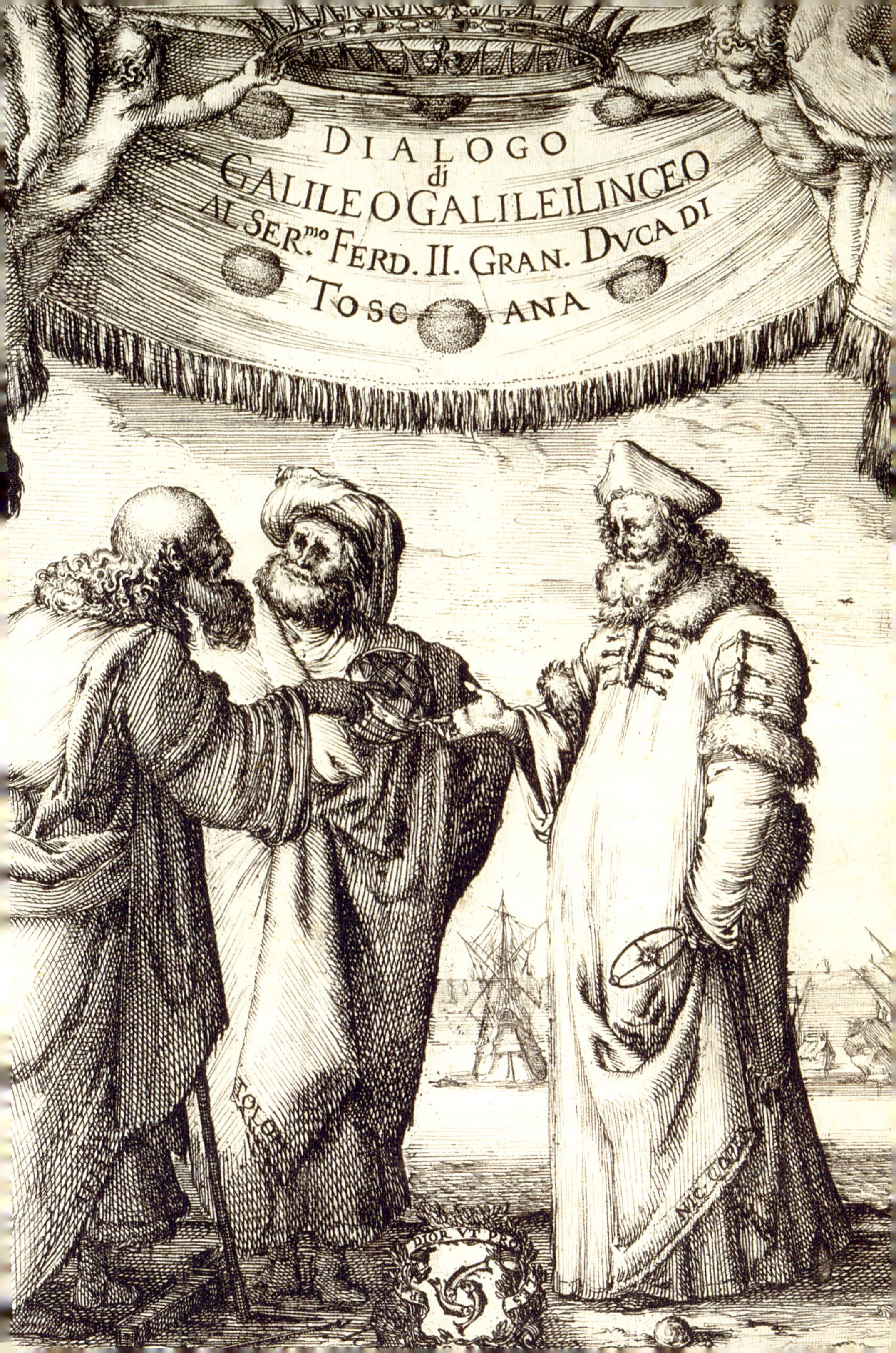
DIALOGO
di
GALILEO GALILEI LINCEO
AL SER.mo FERD. II. GRAN. DVCA DI
TOSCANA

subject of very heated discussion in France, Italy and the Low Countries. The Jesuits, led by Cardinal Bellarmine and echoed by Galileo's patron Pope Urban VIII, tended to favour the idea that the Copernican system was just one hypothesis which might be used to 'save the appearances', or account for observed phenomena. But the fact that it managed to predict the observed phenomena could not be taken as evidence for its truth—different sets of causes could easily lead to the same set of observed phenomena. Galileo put this argument in the mouth of the foolish Aristotelian philosopher Simplicio in his 1632 *Dialogue Concerning the Two Chief World Systems, Ptolemaic and Copernican*, insulting the Pope in the process and almost certainly helping to precipitate his infamous Inquisition trial of 1633, after which he was forced to publicly recant his Copernican beliefs and remain under house arrest for the remainder of his life.

Is mathematics just about measuring and calculating or is it about achieving a deeper level of understanding of the universe? This is a question which was critical to the debates around nature of the early decades of the seventeenth century and which seems to be posed by the *Linder Gallery*. But how much did the Flemish artists of the time know or care about contemporary discussions around the cosmos?

Rubens seems to have had a strong interest in the Copernican debate and a deep curiosity for emerging experimental science. He was interested in Neo-Stoic philosophy (which favoured fluid heavens) and, according to some scholars, portrayed Galileo in his *Mantuan Circle of Friends* of 1602, now in Cologne. The gallery interiors of the Brueghel and Francken studios suggest a high degree of immersion in the scientific culture of the day. Ultimately, a painting is the "deposit of a social relationship", as Michael Baxandall puts it, the material result of a series of decisions made through interactions between artists, patrons and other interlocutors, and the *Linder Gallery* was born through the conversation between at least three people—an unidentified Flemish artist, an Italian mathematician (Muzio Oddi) and a German merchant, Peter Linder.

Is this extraordinary picture a plea for cosmological tolerance or a militant claim for the power of geometry and arithmetic to help us understand the universe? Or is it a reaffirmation of the status quo, reducing mathematics to the humble practice of measurement and calculation? Is it celebrating the dialogue between painting and mathematics or suggesting that painting is dead? Is the male figure intended to be a portrait of Kepler or just a personification of *Disegno*? What process led to the painting's creation? What are we still not seeing in this painted copper surface, this mirror of the political, scientific and artistic ferment of the 1620s?

In looking at the *Linder Gallery*, in delving into its meticulous details, we are gazing into a forgotten world of witchcraft and cosmic harmony, of connoisseurship and iconoclasm, of melancholy and mathematics. Welcome to the conversation!

MICHAEL JOHN GORMAN

JAMES M. BRADBURNE

CONVERSATION AROUND
THE LINDER GALLERY

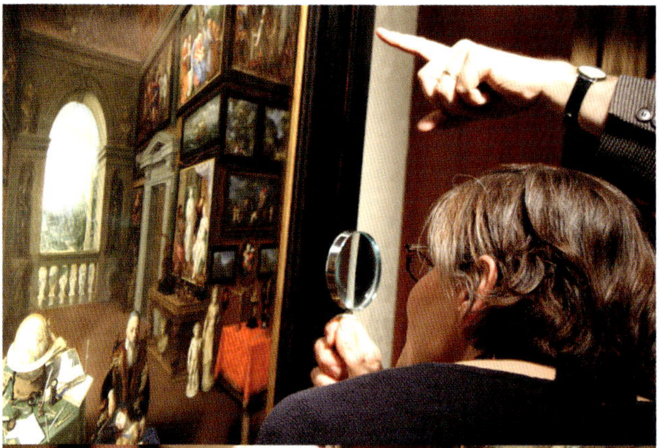
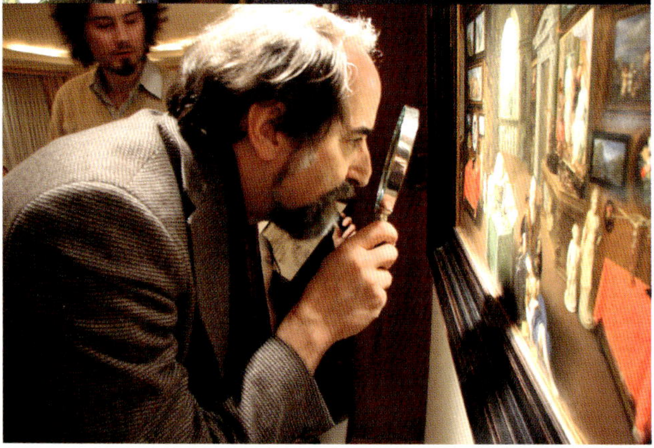

Recorded on 12 December 2008
in the Cordover residence in New York
in the presence of the original painting.

PARTICIPANTS

JAMES M. BRADBURNE is a British-Canadian architect, designer and museum specialist who has designed World's Fair pavilions, science centres, and international art exhibitions. Educated in Canada and England, he developed numerous exhibitions, research projects and symposia for UNESCO, national governments, private foundations and museums worldwide during the course of the past fifteen years. He is currently Director General of the Palazzo Strozzi Foundation, which is responsible for Florence's largest temporary exhibition space, the Palazzo Strozzi.

RONALD H. CORDOVER is an entrepreneur and financial executive. He is presently Chairman and CEO of the Robert Allen Group, the leading US designer and marketer of decorative fabrics and furnishings to the interior design trade. He is an MIT-educated scientist with a PhD in laser physics/electrical engineering. He and his wife Barbara are also art collectors with broad interests including seventeenth-century Flemish art.

MICHAEL JOHN GORMAN is the Founding Director of the Science Gallery at Trinity College Dublin, and has worked extensively in Europe and the US on developing and curating exhibitions and festivals connecting science and art. He is the author of several books including *Buckminster Fuller: Designing for Mobility* (2005). He has been investigating the history of the *Linder Gallery* since 1998 and has written, with Alexander Marr, the first critical study of the painting, '"Others see it yet otherwise": *disegno* and *pictura* in a Flemish gallery interior', *The Burlington Magazine*, no. 1247, 149 (February 2007), 85–91.

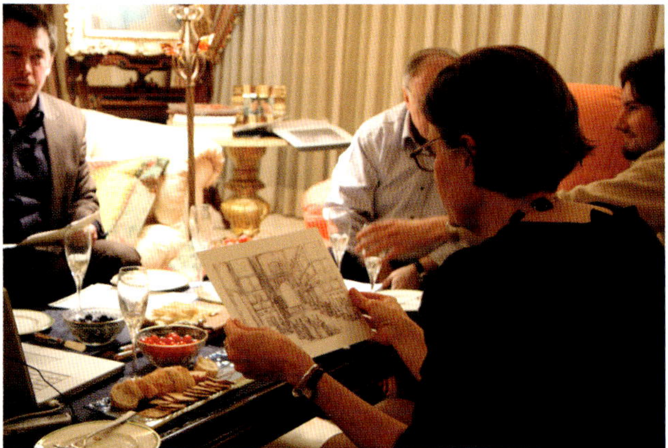

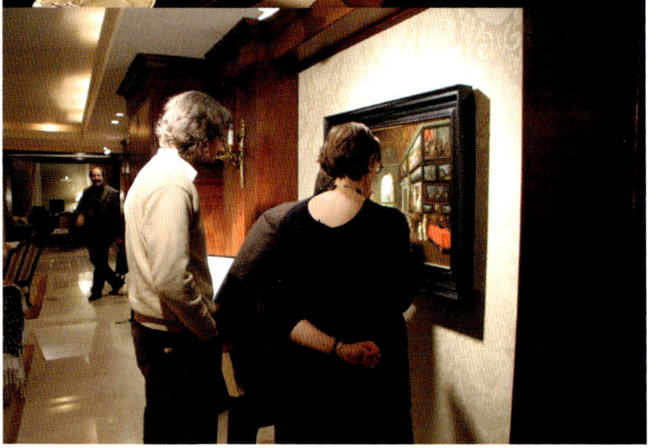

ALEXANDER MARR is a Lecturer in Art History at the University of St Andrews. He has published widely on art and science in early modern Europe and is the co-author, with Michael John Gorman, of the first critical study of the *Linder Gallery*, "Others see it yet otherwise": *disegno* and *pictura* in a Flemish gallery interior' (2007). His monograph on the mathematician-architect Muzio Oddi—*Mathematics and Material Culture in Late Renaissance Italy*—is forthcoming from the University of Chicago Press and he is currently editing a special issue of the *Intellectual History Review* on the 'pictures of collections' genre.

PAMELA H. SMITH is Professor of History at Columbia University, New York, where she teaches early modern European history and the history of science. She is the author of, most recently, *The Body of the Artisan: Art and Experience in the Scientific Revolution* (2004) and the co-editor of *Making Knowledge in Early Modern Europe: Practices, Objects, and Texts, 1400-1800* (with Benjamin Schmidt, 2008). In her present research, she attempts to reconstruct the vernacular knowledge of early modern European metalworkers from a variety of disciplinary perspectives.

LAWRENCE WESCHLER is Director of the New York Institute for the Humanities at New York University, Pulitzer Prize finalist and author of many books including *Mr. Wilson's Cabinet of Wonder* (1995), *Boggs: A Comedy of Values* (1999), *Vermeer in Bosnia: A Reader* (2004) and *Everything That Rises: A Book of Convergences* (2006). Most recently he has published two conversation books, the expanded edition of *Seeing is Forgetting the Name of the Thing Ones Sees*, recording over thirty years of conversations with Robert Irwin, and *True to Life*, recording conversations with David Hockney over twenty-five years.

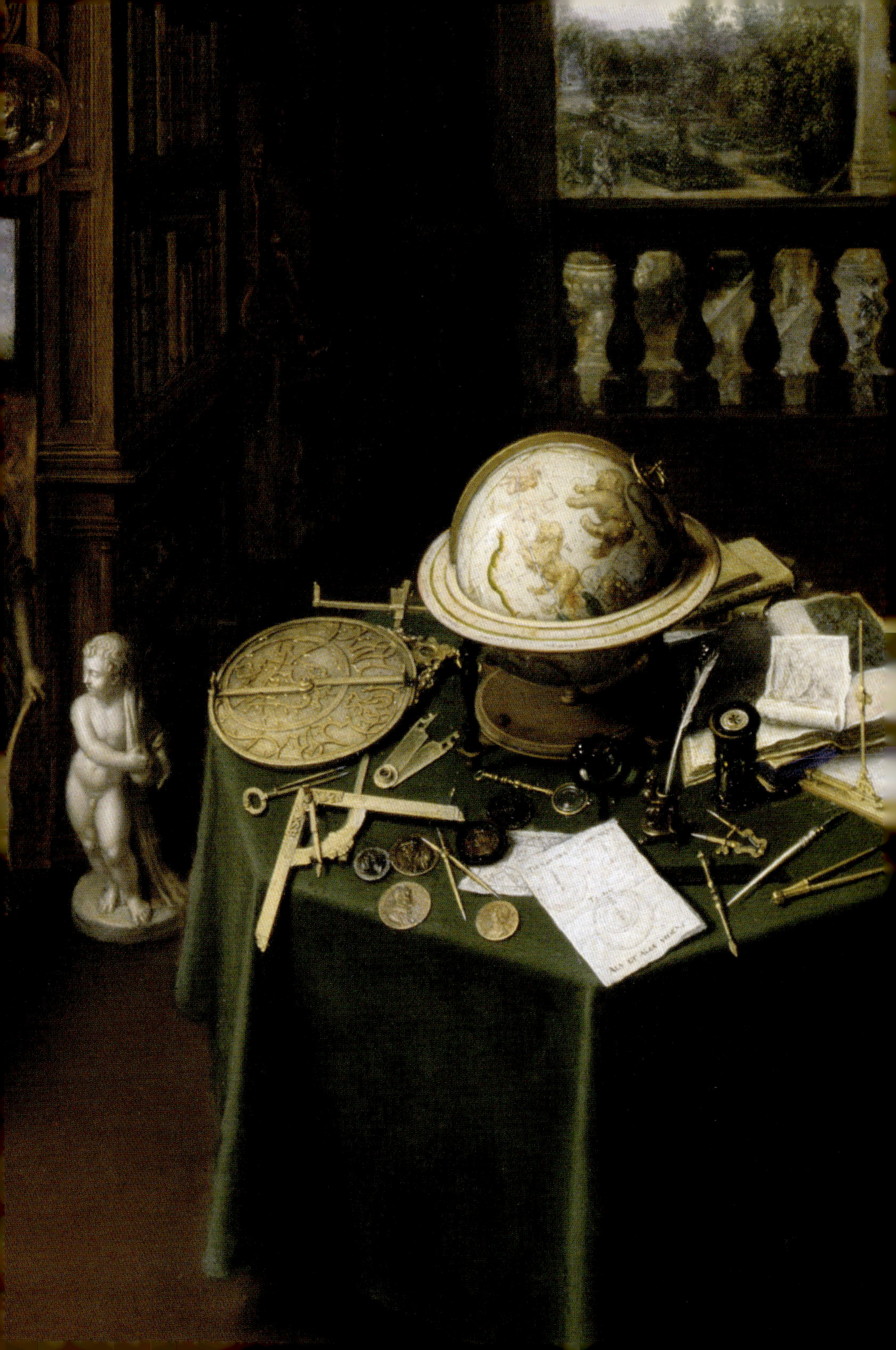

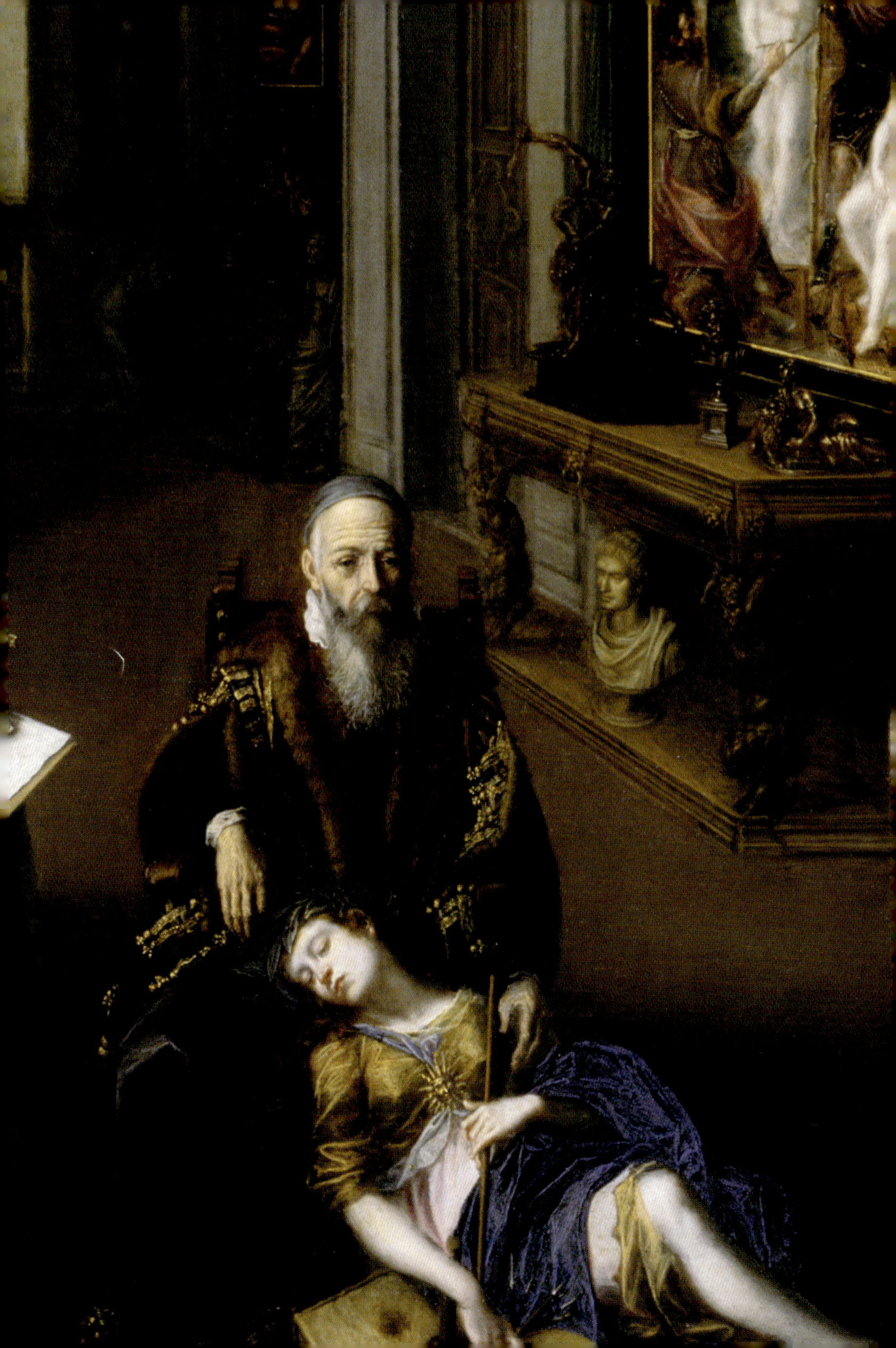

GORMAN | Well, here we are—sitting in front of a painting that was designed almost four hundred years ago as a puzzle. A painting designed to provoke a certain kind of conversation and a certain kind of deciphering. And it's a remarkable painting. It depicts the interior of a picture gallery with a collection of paintings and an assortment of different kinds of sculpture, and also a very impressive selection of astronomical and mathematical instruments, a series of drawings and portrait medals, as well as a pair of figures who occupy a really important place in the foreground—an old man with a young woman reclining on his lap. Ron, what was it about this anonymous and previously unknown painting that drew you to it as a collector?

CORDOVER | The principal initial attraction was the extraordinary execution of the work. Its level of detail is really extraordinary. Its being on a copper substrate allows for that special character. Also the subject matter being treated is one that is so much a part of our personal history in science and the interaction between sci ence and art... And the fact that this painting was unknown. At the time we acquired the painting there was no knowledge, at least published knowledge, of its provenance before the previous owner acquired it in the mid-twentieth century. And it hadn't been attributed. For a painting of this quality to not have been attributed seemed odd.

WESCHLER | How was it presented to you?

CORDOVER | It was the cover lot at a major auction in one of the New York auction houses—at Christie's. And it was presented as an "important Flemish gallery painting".

WESCHLER | Among the lot itself or in the auction were there other gallery paintings or Flemish paintings?

CORDOVER | It was the only gallery painting in their normal Old Master paintings sale. But the collector, Thomas Mellon Evans, was a collector of broad interests and had his collection posthumously redistributed through a series of sales that included important Impressionist and post-Impressionist objects. But this particular painting was fascinating because so little was known about it and yet it was such an obviously special object.

GORMAN | Do you think that your background in physics affected your engagement with the painting?

CORDOVER | Well, there's no doubt about that because in the centre of the picture, thrust out so that you can't miss the artist's intent, is a wonderful summary of the body of knowledge about the movement of the planets around the Sun.

GORMAN | You said that when you got it there was very little known about the provenance. What have you learned since about the story of the painting?

CORDOVER | Well, we were fortunate that we knew where the owner had ac-

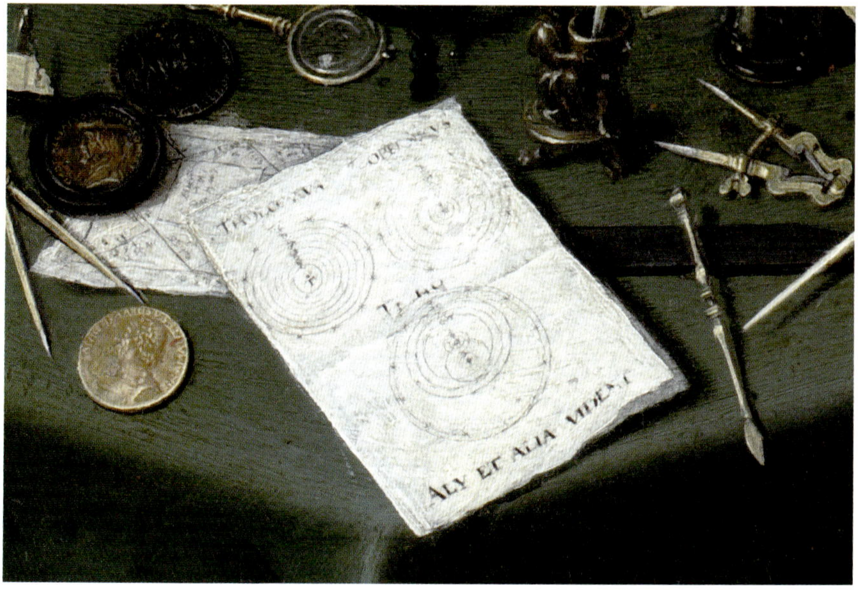

quired the painting, and in the old archives of that art gallery we found a series of letters from the Rothschild family that defined the painting as a Jan Brueghel the Elder object in the Rothschild collection in Vienna. It was thrilling to make that discovery, but also quite disturbing, because we knew that much of the Vienna collection had been confiscated by the Nazis at the beginning of the war. Subsequent research in the Rothschild Archive in London has shown that this, along with five or six other objects, was granted an export license after the material was rescued by the Allies at the end of the war, and before the Austrian government took the remainder of the rescued Rothschild collection into the Austrian national museums.

GORMAN | Right. When the Nazis took the painting from the Rothschilds, they kept it in a salt mine in Salzkammergut where they stored a huge amount of the paintings they had taken from museums and private citizens. And in the Nazi inventory in the Rothschild Archive in London there's a little red tick beside this picture. And the little red tick meant that this was intended by Hitler for the Führermuseum in Linz, which was Speer's enormous museum project. So that was where it was in-

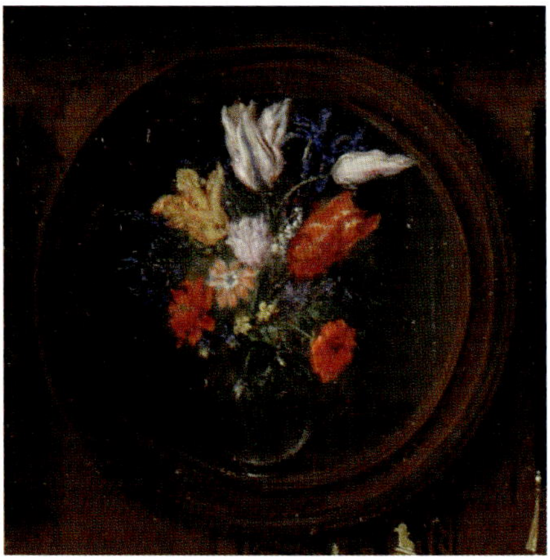

tended to go except for the Allies coming and restoring it to the Rothschild family, from whence it made its way to the US. But now I'd like to go back into the world of the seventeenth century, the moment when this picture was created, and I'd like to turn to you, Pamela. What's this picture saying about material culture and about how people collected objects in the seventeenth century?

SMITH | I think that if you look at the types of objects, there were three categories in the Hapsburg collections: *scientifica*, *naturalia* and *artificialia*. And that was fairly standard for collections at the time. The scientific instruments, the celestial globe and the surveying instruments—all of those kinds of instruments, even if we now think of them as the applied arts or the applied sciences—were *scientifica*. Then, in terms of *naturalia*, there are no *naturalia* there except the painted *naturalia*. So the flowers, mainly. But really that's about it for *naturalia*.

GORMAN | Except the fruit in the top left.

WESCHLER | Paintings of fruit. Which are, by the way, paintings of paintings of fruit. Though, ontologically, they have the same status, paradoxically speaking.

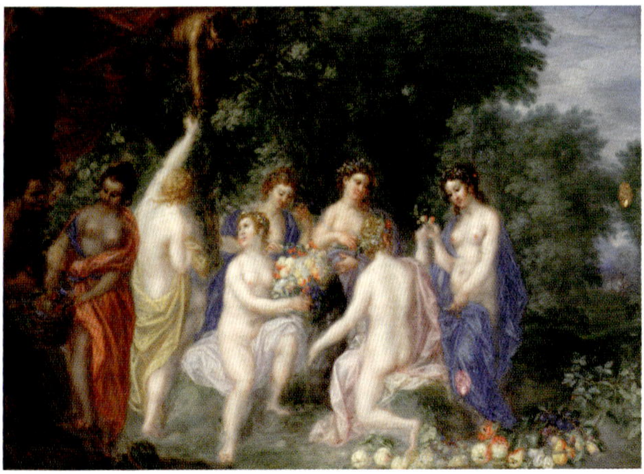

SMITH | Exactly. Although it's actually one of the recommendations stated in some books of painting in the seventeenth century. For example, Gérard de Lairesse, late seventeenth century, says that you should have paintings of flowers and fruits and animals in order to substitute for the real thing. So we have *naturalia* there in a sense, although it's not as obvious as it would have been in an actual Kunstkammer.

MARR | And of course we have the natural world beyond the gallery.

GORMAN | Right, the garden. But what's this picture saying about our desire to understand the natural world?

SMITH | Well... First of all I think one of the guiding keys is mathematics. I mean, that's clear from the whole picture. I think the crucial point is this intersection between what we think of as science and art at the time. They were really seen as part of the same endeavour.

CORDOVER | I think that's a critical observation. We have to see this with seventeenth-century eyes, not twenty-first-century eyes.

WESCHLER | It's around now that they begin to delaminate from each other, and the way it continues to this day is in the way we talk about 'Arts and Sciences'. A liberal education is Arts and Sciences. In fact things have descended to the point where nowadays, in newspapers, it's Arts and Leisure. But in those days there was absolutely no difference between the two—Michelangelo is a scientist and an artist, and that's the same thing. And it only begins to fall away around that time.

CORDOVER | Yes. Michael John made the observation that this painting is intended at least partially as a puzzle painting, in the sense that there are a lot of clues to deeper meanings, but there is also within this painting, especially on that centre table, a statement of cumulative knowledge about the real world. You know, it's not a puzzle at all to assert observations about measuring.

GORMAN | I think that's important. There's a lot here about the collection of empirical data about the natural world, and particularly the positions and movements of the stars and planets.

CORDOVER | Which is why this painting was exciting for us. It's because this, in our minds, was the first moment when pragmatic experimentation and observation of the real world began to joust with other philosophy, other reflection—even religious reflection—about what our universe really represents.

WESCHLER | Taking it one step further, I would also argue that this is the mo-

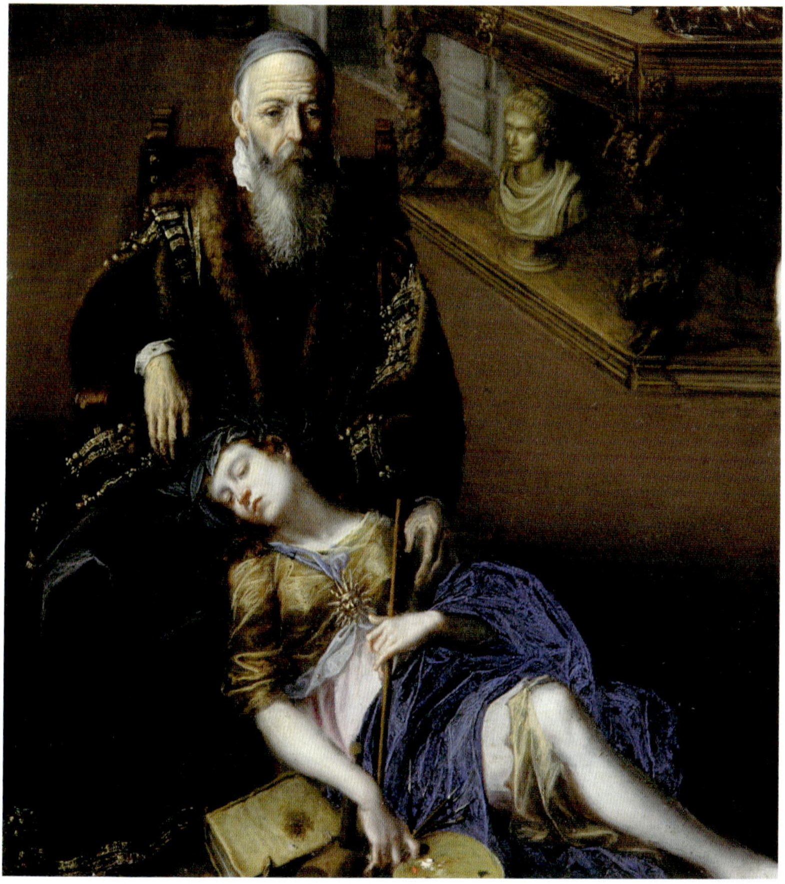

ment when art starts being art as opposed to being science. For all of your claims that the old man and the woman should be read principally iconographically—that they represent this or they represent that and so forth—there *are* times when a cigar is only a cigar, and there is something quite moving about that girl asleep on the lap of that old man—something crying out to be seen not just as a symbol, but as a relationship between two human beings. Thirty years from here you'll be getting Vermeer with his insistence on the substantive presence of the sitter as a person and not as a genre and not as a symbol, not as a puzzle—not as anything other than *a person*. And I think that's already beginning to happen in a time like this.

GORMAN | Alex, where does this picture sit in the genre of pictures of galleries?

MARR | Well, it's interesting in light of the point Ren made about this being a moment when art becomes art, because this genre of gallery interiors is precisely about that. It's about the moment at which art appreciation becomes a recognized phenomenon, a recognized and accepted thing to do. This particular genre of pictures of collections emerged at exactly the same time that the *liefhebbers van schilderen*—the so called 'lovers of painting'—were first admitted as a category of member to the Guild of St Luke, the artist's guild in Antwerp. So it's a recognition that art isn't just about making. It's about appreciating and evaluating and discussing. And that's really what this genre meant to be about. It's showing images that cognoscenti could stand in front of discussing differences between style, differences between manner, differences between treatment of themes.

CORDOVER | While I agree completely that this genre of painting—Flemish gallery painting—is precisely as you've described it, this work has such striking differences from essentially all the other Flemish gallery paintings that we've looked at because it doesn't have cognoscenti within it and it isn't principally about a particular collection of either art objects or other kinds of objects—musical instruments or *artificialia* or *naturalia* of any sort. It's about knowledge.

MARR | I would agree with that, but I think this is essentially a hybrid. Yes, you're quite right—there are no cognoscenti actually in the picture. I don't think that that means that this isn't an image that should be thought about in terms of cognoscenti and appreciation.

WESCHLER | This reminds me of David Hockney, who has recently taken to painting these landscapes which he calls his "figure paintings" And if you say to him,

"But where are the figures?" he replies, "*You* are the figure." In the same way, in answer to the question of where are the cognoscenti—*you* are the cognoscenti.

MARR | And it is also about knowledge—but it's about lots of different kinds of knowledge. I think there's no doubt that knowledge of the cosmos and how the world fits together is a key theme. But I think that we shouldn't forget that this is also about recognition of the styles of different painters. So I think it's about—going back to arts and sciences and how they're all mixed up—I think this is an image about different kinds of knowledge.

GORMAN | You could say the picture is a machine to produce cognoscenti.

SMITH | But mathematical knowledge predominates in the painting. I mean the mathematics of measurement… Don't you think?

MARR | Well, yes and no. If we were to count up the number of mathematical objects and the number of art objects I think we'd probably find that the art objects outnumber the mathematical ones.

GORMAN | There's a question of prominence though. The mathematical and astronomical instruments are especially emphasized…

WESCHLER | They're actually pressed forward and lit up in a particular way.

MARR | It's notable, isn't it, if we agree—as I'm sure we can—that a key element of this painting is about mathematical knowledge, that the setting in which this is being presented is a gallery interior. Or maybe rather we should call it a study or a studio or a Kunstkammer—whatever we want to use. This is a room in which lots of different kinds of objects and individuals meet. Just like here.

CORDOVER | Well, I attribute that to the fact that the object itself is an extraordinary work of art that comes out of the studio of an important Northern European painting family that had, along with others, created a series of these kinds of objects. So that even though the fundamental idea for the theme of this painting is the assertion about the knowledge of the cosmos and the exploration of that knowledge, it's put into the setting of this pure art form.

GORMAN | In the Brueghel–Rubens *Five Senses* series, surely linked to this picture, you have for example the *Sense of Sight* with a very different gallery space with dogs, parrots and monkeys, and there are many gallery interiors that have animals cavorting through the gallery, and it's a very different kind of space from this picture. And then, I suppose, at the other end of the spectrum one has the later gallery interiors of people like David Teniers the Younger, who represented existing collections where you have no mathematical instruments, but you have people—and you have representations of real pictures. So where does this picture fit in that sequence?

MARR | Maybe it would be helpful just to give a very quick overview of the genre?

WESCHLER | Yes, so tell us the history of this genre. Where does it come from?

MARR | Well, the gallery interior is invented in the second decade of the seventeenth century. We're not sure whether it was by the Francken studio or the Brueghel studio, but they're amongst the earliest developers—key Antwerp cabinet painters. There are several broad categories of picture, all of which overlap so that the boundaries between them aren't clear-cut. In essence you have what are called allegorical cabinets. Michael John referred to the famous Brueghel–Rubens' *Sense of Sight*. That would be an example of a painting in which very obviously allegorical figures are inserted into a gallery setting for some kind of emblematic or narrative purpose.

GORMAN | Which this picture would also represent.

MARR | Right. Then you have so-called portraits of collections, which are images that seem to record, with varying degrees of faithfulness, actual collections that existed in the period. The best examples are the later ones from the 1650s by Teniers of Archduke Leopold Wilhelm's famous collection, of which there are numerous examples.

WESCHLER | And why would that have been painted, by the way? Was that because when he went travelling he wanted to be able to show off his collection?

MARR | Most frequently they were meant as gifts. Partially they're records. Some of the Teniers paintings were later engraved as a sort of a proto-catalogue. But often they were gifts.

CORDOVER | Also, to show off.

MARR | Yes.

WESCHLER | Is it like my having a picture of my daughter in my wallet so I can show you?

MARR | It's more that, if a prince is visiting a member of the aristocracy in a neighbouring state, he might bring this as a gift. Which is a nice present, but also displays the prince's own magnificence. And in a sense it's about fashioning himself as a great patron and connoisseur.

WESCHLER | Or as John Berger would have said, the upsurge of oil painting and the upsurge of property are simultaneous.

MARR | Yes, very much so. Then there is the other type of gallery interior which is neither a portrait of the collection nor is it very obviously allegorical. But it seems to show connoisseurs within a cabinet setting discussing various kinds of object. It will always include pictures. Almost always include statuary. Frequently it will include other kinds of *artificialia* such as medals, sometimes scientific instruments. Also, sometimes, *naturalia*. And it is frequently the case that gallery interiors are collaborative works, by many different painters, with the identification of the hands of the different authors being part of the pleasure of owning such a work.

CORDOVER | But the objects that are identifiable tend to be the scientific instruments and not the paintings.

MARR | Yes. The pictures within the pictures that you see in these genres are sometimes identifiable as existing pictures that we can see hanging in galleries today. Sometimes they're not. But that was part of the fun—that these galleries included pastiches as well as real images.

WESCHLER | Oil is very good at portraying stuff you have, as opposed to fresco, which might be good at portraying God or the Crucifixion and things like that.

MARR | This is very much about consumerism, there's no doubt about that. Antwerp is a great trading centre. It's a massive centre of commerce and it has a large middle class. So that's an important point to make about the evolution of this genre in that particular place at that particular time. It's to do with trade and mercantile aspiration.

GORMAN | But Antwerp was falling into decay at this time though, as the Dutch were really taking off. Amsterdam was emerging as the new centre of commerce. After the Dutch blockaded the river Scheldt, Antwerp began more and more to focus on luxury goods including paintings and jewellery. Perhaps it's more about commerce past than future.

WESCHLER | For all of the stuff you're saying on the connoisseurship side of it, I do think that this particular genre is haunted by surfeit. There's a 'too-muchness' to it. You would think that somebody who knew how to wield that mathematical stuff would know how to make a table that was in perspective with the rest of the objects, but this one conspicuously isn't—it is conspicuously out of perspective. This allows me to have two ideas about what might be going on. One of them is that although the green table is effectively a circle, done like that it becomes an ellipse, which is itself a comment on Kepler.

GORMAN | Isn't it an octagon?

WESCHLER | Well, yeah, it is an octagon, but it suggests a circle, though in fact, if you look at it closer, it inscribes an ellipse as opposed to a circle, so that's one possibility. But the other thing that seems to me more interesting is it is suggesting a sort of Land of Cockaigne. Peter Bruegel the Elder famously does a painting in which there's a table exactly like that, at an odd upturned angle, and the theme of that painting is precisely this too-muchness. I mention that because the contemporary painter Vincent Desiderio recently did a painting which he calls *Cockaigne* and which also conspicuously references the Bruegel. As in the Bruegel, Desiderio's is a painting of after a feast, the feast is over, and scattered all about the floor are hundreds and hundreds of art books. Just too much… And there is that feeling in this painting as well to me.

CORDOVER | There's an extraordinary element in the observation about the table being out of perspective. It is a deliberate 'error', because the study drawing from which the painting obviously comes has that same table in perfect perspective. Making the table the centre of attraction is accomplished by placing the objects on it in ways that call attention to them, and by creating the odd perspective for further emphasis.

WESCHLER | It's thrust in your lap and it's about to fall into your lap.

SMITH | Yeah, but to me it highlights it and says this is the key to the painting. The Land of Cockaigne is fascinating. I never would have come up with that.

WESCHLER | I basically am to art history what Pluto is to the solar system. I'm not even sure whether I'm a planet.

SMITH | But has any work been done on whether there are any underdrawings?

GORMAN | Yes. There's a very clear perspective underdrawing. We have some of the infrared studies here.

SMITH | Oh great. And so, does it show other objects? Because my thought is actually that there is far less copiousness in this picture than in other gallery interiors.

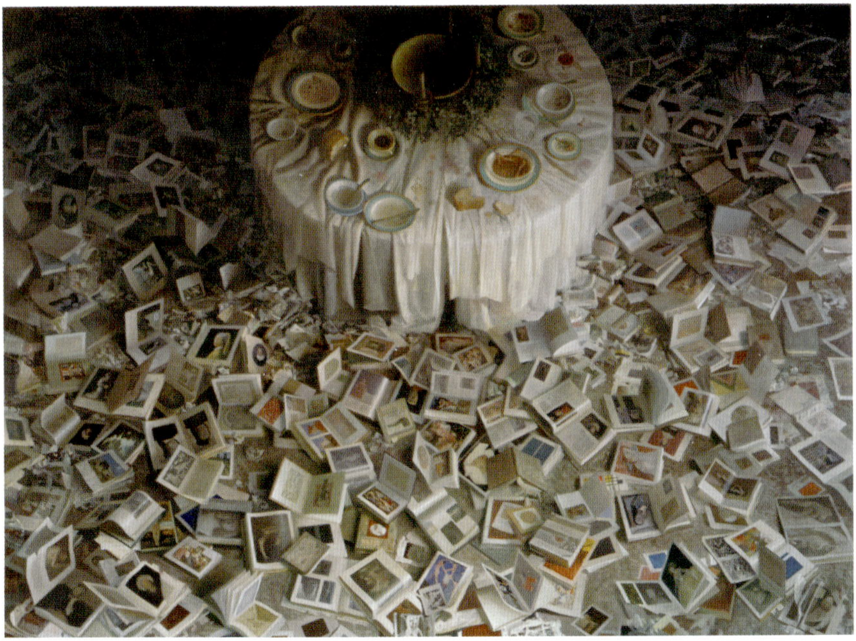

Vincent Desiderio (Philadelphia, Penn. 1955–), *Cockaigne*, 1993–2003, oil on canvas, 284.2 x 389.6 cm. New York, Marlborough Gallery.

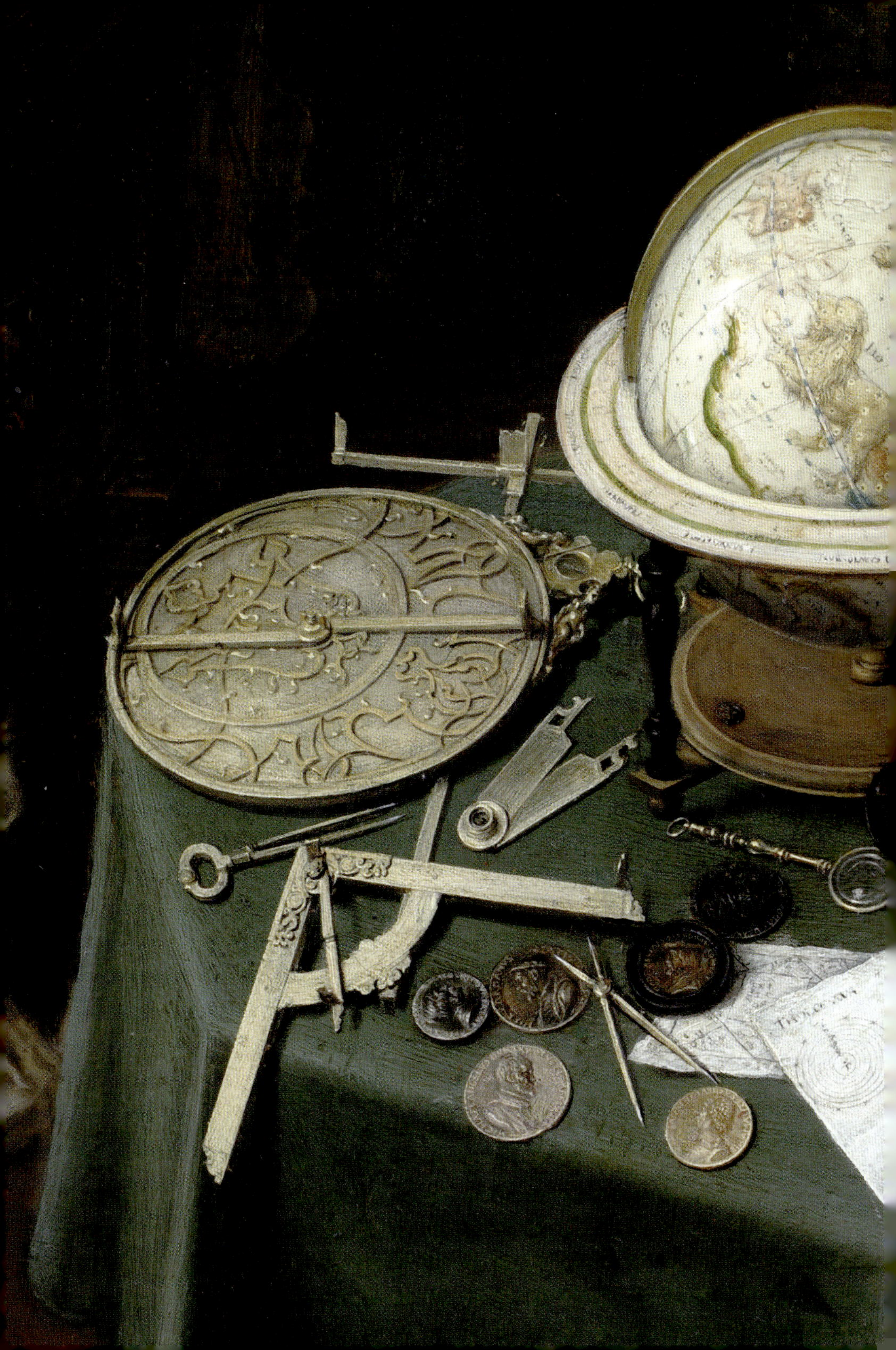

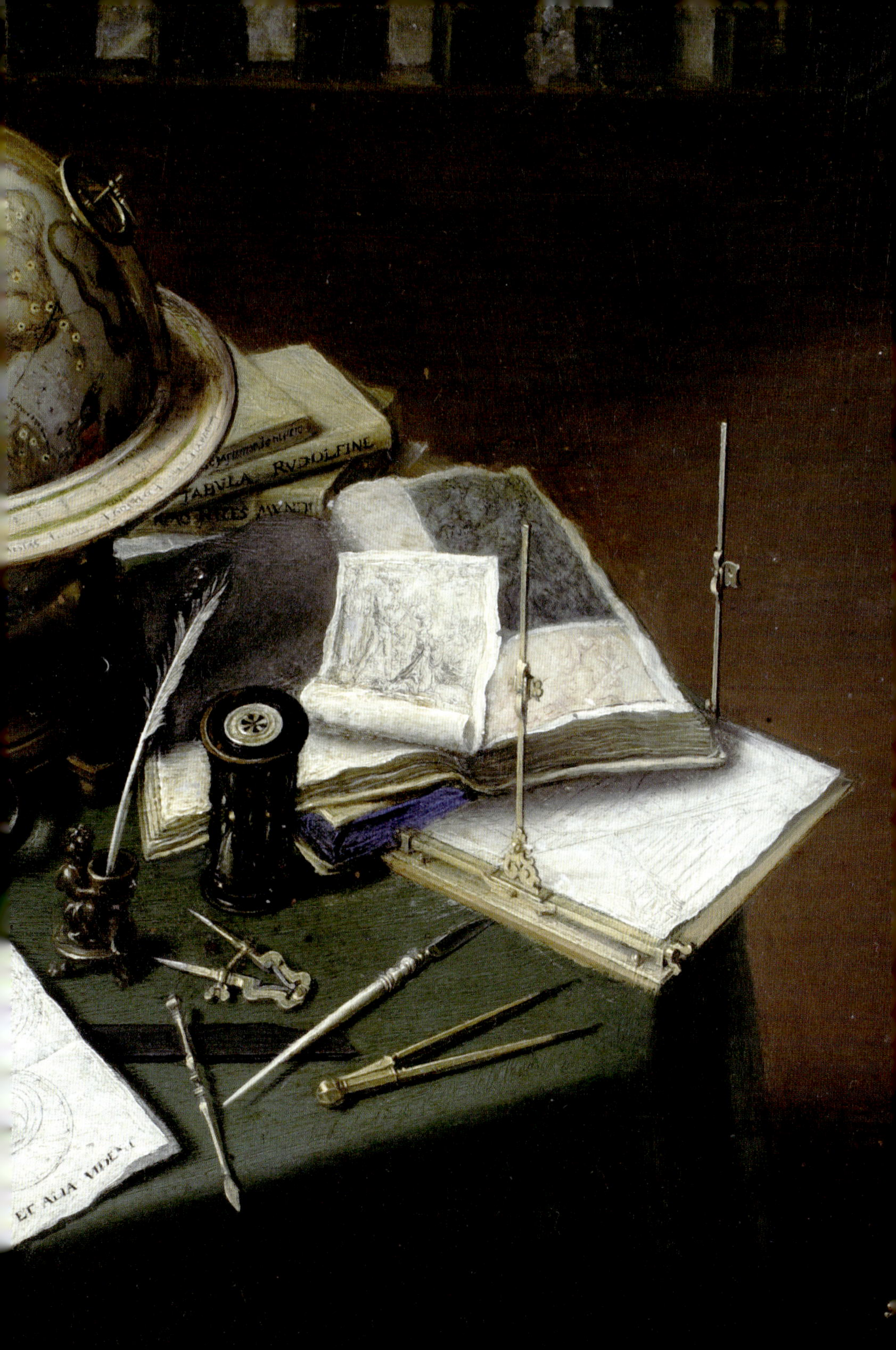
TABVLA RVDOLFINE

GORMAN | The *Sense of Sight*, for example.

SMITH | Yes. Exactly. There's much less exotica, much less that is supposed to seduce the eye.

GORMAN | Less voluptuousness, really.

SMITH | Exactly. And the nude Campaspe is absent from the study drawing, and if there is this woman showing her leg there, she's still very clothed. And she's allegorical. She's clearly allegorical.

CORDOVER | And she's clearly Pictura. She's got paintbrushes and a palette.

SMITH | So she's no longer just a voluptuous nude. But, you know, it really strikes me just how little this contains of all of the sensory and sensual elements that you see in other gallery interiors.

WESCHLER | My masters in art were both literature professors. Edward Snow and Harry Berger. In his great book on Peter Bruegel the Elder, *Inside Bruegel: The Play of Images in Children's Games*, in which he chronicles fifteen years spent looking at one painting, Snow makes the rather astonishing assertion that the only mind as capacious as Bruegel's in the history of the world might be Shakespeare's. Now, I don't want to make such a sort of claim for this painting, but it is interesting that a painting can contain both extremes—in the way that a Shakespeare play can be about both overabundance and starvation. I think it's T.S. Eliot who at one point has that line about history gives, but she gives "with such supple confusions that the giving famishes the craving". And I think it's worth imagining how in this period, not only have art and science not yet fully delaminated, but neither have literature and art fully separated. This is a few years after Shakespeare, right? A mind like Shakespeare's was capable—and you could assume that an audience would likewise be capable—of getting both starvation and overabundance, and I would think that you could look at paintings that way too. Such that a lot of these things that are paradoxical or puzzle-like in this image are finally about human nature. About what it is to be human and to be filled with contradiction.

SMITH | And what humans can know.

WESCHLER | Do art historians read Edward Snow's book on Bruegel? I think it's one of the great books of all time and I really recommend it. It's fifteen years of looking at one painting. And as with what I assume will be the case with this book, it has dozens, hundreds of details in it. At the time I read his book, I asked Snow, "But

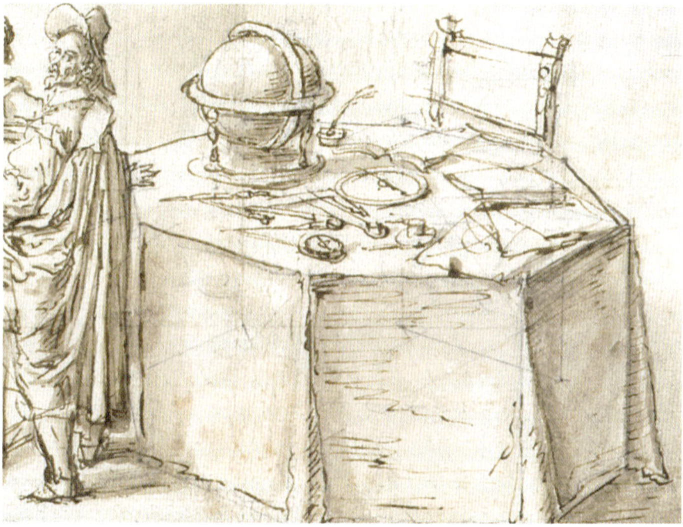

A Picture Gallery, Windsor Castle, RL 12983, detail.

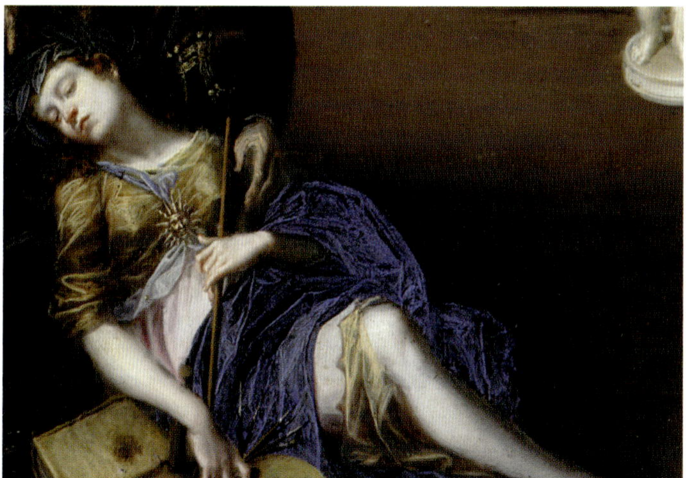

isn't this a peculiarly anachronistic way of looking at a painting, since you can only talk about the painting this way when you can publish your analysis along with all these details, and they couldn't have done that in the Old Days. Isn't this just the way we look at things nowadays because we happen to have the technology to be able to take details, blow them up and so forth and so on." And he blew up at me and he said, "On the contrary. Understand," and he was talking about Bruegel the Elder at this point, "that this is probably one of the most brilliant men who ever lived. He is working on a single painting over let's say six months, eight months, a year. He is having ideas about things he painted yesterday that he is now reconfiguring over here. This is a man much smarter than you or me and he's having all these kinds of things going on, detailed thoughts about this particular location and how it balances against that and so forth. He's containing multitudes. And he's doing it for somebody who will eventually own the canvas and place it in his study or over his desk or something, in a time when he doesn't have TV sets or posters or magazines. It is intended to be looked at for fifty years and every day have something new be seen in it. In fact, what is anachronistic is to come upon that painting in Vienna and spend ten minutes looking at it in a room crammed with twenty similar paintings. That's not the way you're supposed to look at it."

CORDOVER | Earlier, we were discussing the complexities of this painting and the fact that there may be a number of hidden meanings, you know, a surfeit of big ideas to explore over a lifetime of viewing. It's very clear that at the centre of the painting there is this oddly presented table, and in the most prominent place in the painting, thrust forward on the table, there is a drawing that is just shouting at you, 'this is what I'm about'. That drawing illustrates the debate that was going on between scientists at the beginning of the seventeenth century, following Copernicus' assertions in the middle of the sixteenth century that Ptolemy had it wrong placing the Earth at the centre of the solar system and the Sun and other planets revolving around the Earth. Copernicus, of course, placed the Sun at the centre. Finally there is a third suggestion here by Tycho Brahe, the astronomer to the Hapsburg emperor until the end of the sixteenth century. He asserted that there is another solution to cosmology that allows the Earth its prominence at the centre; but it requires a tortured mathematics that isn't persuasive to anyone but perhaps those with a fundamentalist religious bias.

WESCHLER | So the order, by the way, historically is Ptolemy, Copernicus, and then Tycho Brahe comes after that?

CORDOVER | Correct.

WESCHLER | That reminds me of the phenomenon of how belief breaks down. I have a footnote in *Mr. Wilson's Cabinet of Wonder* where I talk about how my daughter believed in Santa Claus, then stopped believing in Santa Claus and then the following year had a more elaborate, more insanely baroque belief in Santa Claus than ever before. And if you think about it, that's the same thing that happens with the Age of Marvel. In order to get to America in 1492, there had to be an incredible amount of positivist mathematical knowledge of the sky, the stars and the solar system. But then what was discovered in America was so weird as to sanction all sorts of belief in things that had previously been debunked. I mean, if narwhal tusks are real, why couldn't there be unicorn horns? And if parrot feathers were in fact all those weird colours, why couldn't all those things that Marco Polo or Mandeville

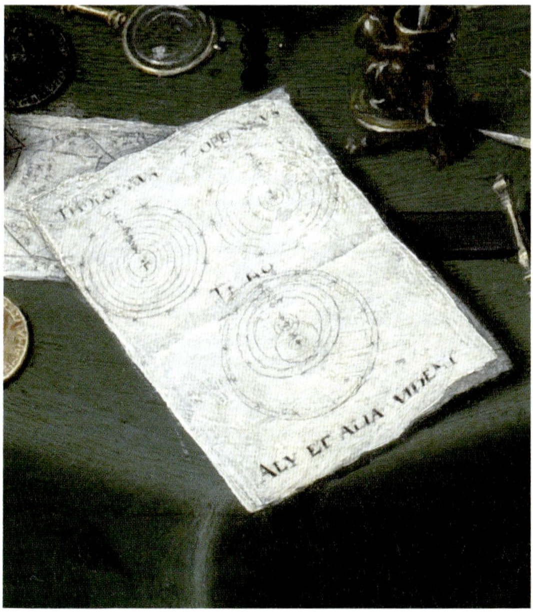

talked about be true? Such that after you had had this kind of positivist upsurge to be able to get to America, you had this kind of backslop of marvel. In the same way, maybe, you go from Ptolemaic nonsense to Copernicanism and then you have this backslop to Tycho Brahe.

GORMAN | I'd like to just step back from the detail a bit and look at the picture as a whole. First of all, the space—is there anything we can say about this architectural space? You've got this T-shaped floor plan, these windows, this closed door… Do we think this is a real or imaginary place?

CORDOVER | It's hard to imagine that it could be completely real, because I don't think you'd have openings to the outside putting at risk all of these fragile and wonderful works of art.

WESCHLER | And books.

GORMAN | Right, but is this meant to be a completely imaginary place, do we think? Or is it a mixture in some way?

MARR | Well, I would plump possibly for a mixture. The simple answer is that we don't know, as we'll see when we get into the details. But certainly gallery spaces similar to this, like this, existed, of this sort of scale with roughly this kind of arrangement. Frequently they did have vistas, admittedly not open and not on gardens, but you commonly have vistas out into the surrounding landscape. It's part of the *paragone*, including the argument between art and nature that was so common at the time.

GORMAN | Well, let's look at this landscape that we see through this open space over this balustrade. Is this a Flemish garden or an Italian garden?

SMITH | Or Roman, with the obelisk.

GORMAN | What is the obelisk there for?

WESCHLER | The obelisks start coming into Rome out of Egypt in many cases. Athanasius Kircher is describing them because he thinks he's figured out how to decipher them.

MARR | It's quite a common feature of Flemish architecture. Hans Vredeman de Vries, the famous Flemish painter who is also a designer of perspectival interiors like this, uses obelisks very frequently in around 1600. So it's an antique motif which crops up in all sorts of places, it just means antiquity.

GORMAN | Or does it? Could it be a reference to astronomy? I mean, if one looks at Christoph Scheiner's books you have the *Rosa Ursina* where he's using obelisks almost as a solar observational instrument. So could the obelisk be an extension of the array of instruments?

MARR | Possibly, and it looks like a gnomon, which is the needle that casts the shadow in sundials. But it's also just a common garden feature.

SMITH | But it also refers to the project of raising the obelisks, which would also be referring to the potential of machines and mathematics.

MARR | I don't know. It's such a common object—just like seeing a column in many ways.

CORDOVER | In looking at this garden, it has a very formal set of elements associated with it, with circular designs, plantings, and yet in the foreground and on the right side of the garden there is a wildness to it that evokes some of the great landscapes of Jan Brueghel.

WESCHLER | I agree with you. If you take your hand and cover that, you're in Eden, you're in Arcadia. It's only there that you see the other.

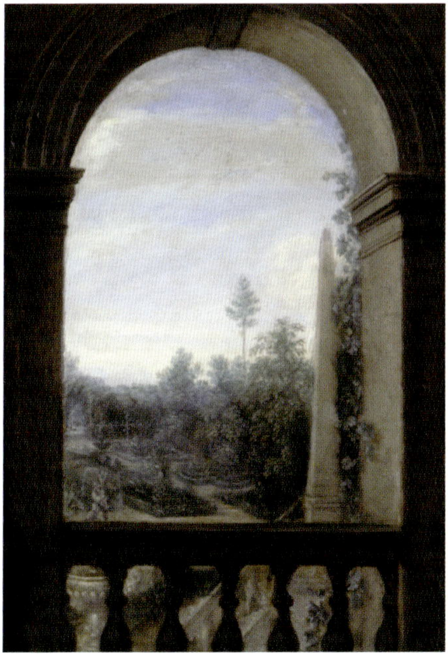

SMITH | But you know what I find curious—there's Rome in the further ground of the painting, there's antiquity, with the arch, the entablature, the obelisk and the open air, and here in the foreground it's Germany with the Linder arms. It looks like a kind of dialogue between Italy and Northern Europe, or Flanders.

CORDOVER | Just look at this. [*He is holding a painting removed from the wall of an adjacent room.*]

GORMAN | This is a Jan Brueghel the Elder landscape.

SMITH | Oh yeah, no question. Yeah. Wow.

WESCHLER | Not everyone can pull rank like this.

GORMAN | I always wonder, are the leaves done the same way in this?

CORDOVER | They are. They are. If you look closely at the singular way he's painted the leaves…

SMITH | The only thing is the vine—the grapevine here.

GORMAN | What about the fountain? Now the fountain, if one looks at it very

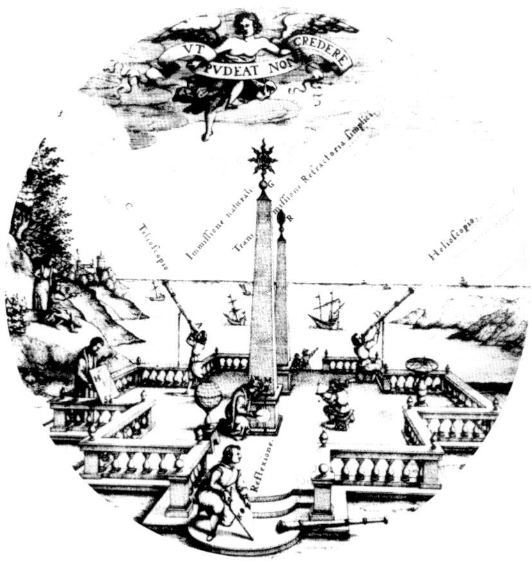

DANIEL WIDMAN, DIFFERENT METHODS OF OBSERVING SUNSPOTS, DETAIL FROM CHRISTOPH SCHEINER, *Rosa Ursina, sive Sol*, BRACCIANO, ANDREA FEO, 1626–30, P. 295, PLATE LVI. FLORENCE, BIBLIOTECA NAZIONALE CENTRALE DI FIRENZE, PALAT. 8.9.7.19.

closely, actually has a putto on top of it playing the bagpipes with water spurting out of the bagpipes, and it's extremely similar to a fountain in Prague—the singing fountain in front of the Belvedere, the building where Kepler and Tycho Brahe carried out their observations. Could that be perhaps a reference to Prague?

CORDOVER | Ursula Härting when she first looked at this painting immediately said that.

GORMAN | Really?

CORDOVER | Yes, she wanted to borrow the painting for a show on gardens, European gardens, and she immediately asserted that it looked like the Belvedere palace garden in Prague.

GORMAN | The fountain is ultimately a hydraulic mechanical instrument, or a machine, so again it is a tantalizing question in the painting whether things are specific or generic—someone could think of the fountain as also representing the mathematical and mechanical arts.

SMITH | As with the obelisk.

WESCHLER | Or being a token for the bounty of nature.

GORMAN | Can we come back indoors now? What is that shadowy object in the right background, James?

BRADBURNE | It is very shadowy indeed, but it is unambiguously a *perpetuum mobile* based on the design by the Dutch inventor Cornelis Drebbel.

GORMAN | What's it doing in the painting?

BRADBURNE | That is a good question. When Drebbel first presented one to King James I/VI in London in 1608, it was considered a wonder of the first order, with its water roiling ceaselessly in the glass tube surrounding a central globe. Despite the darkness, the *perpetuum mobile* is painted with the same incredible attention to detail as the instruments in the foreground—you can even make out the outline of the clock dial in its centre and the globe on top that showed the phases of the Moon.

GORMAN | How did it work?

BRADBURNE | Exactly how it worked is still a matter of some conjecture, as not a single example has come down to us. However, as early as 1612, one of Galileo's contacts in Brussels, Daniele Antonini, made a working model of the instrument and conjectured that the motion of the water was due to changes in air temperature. It seems that the instrument was a thermoscope, and worked due to a combination of

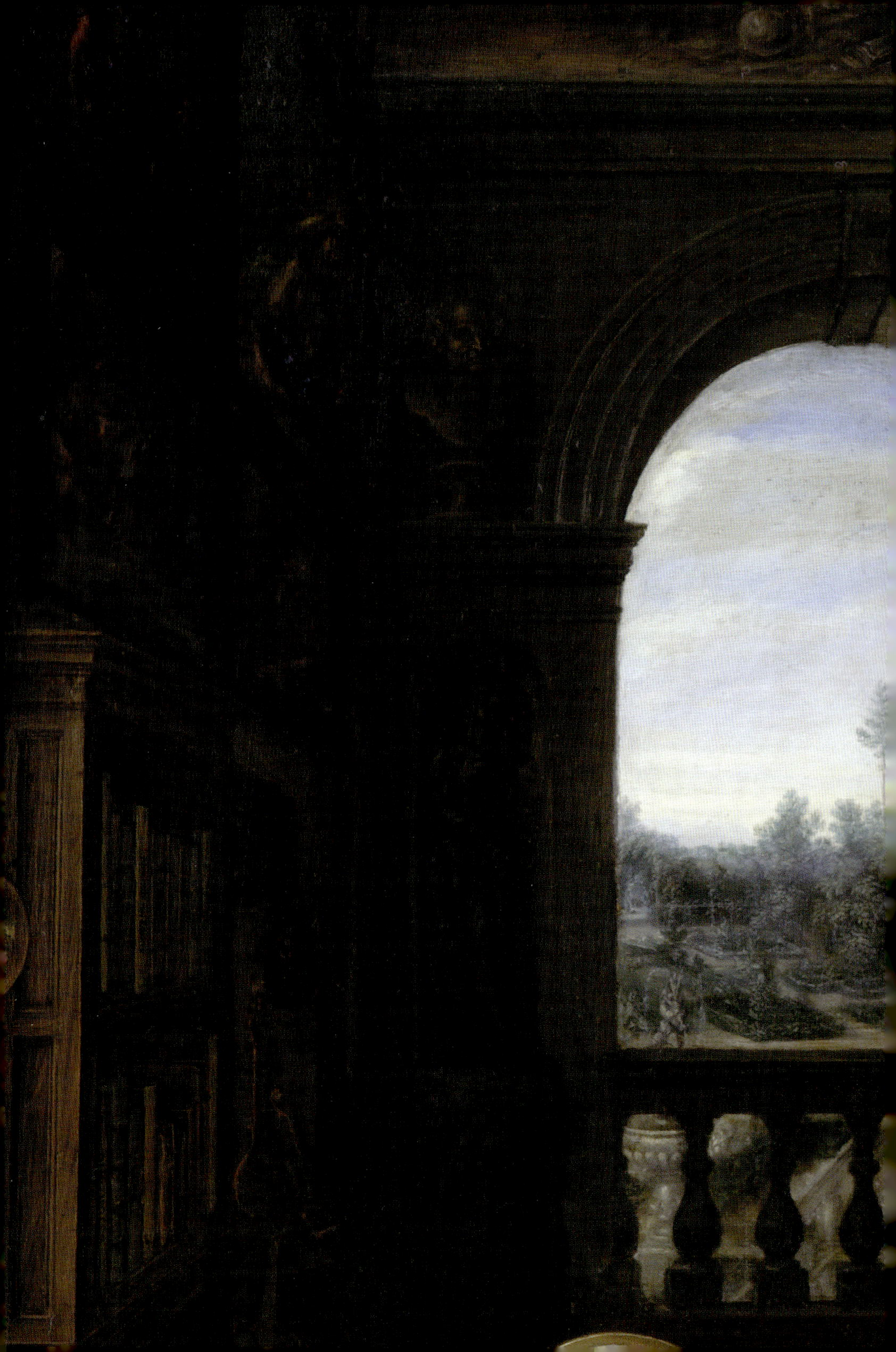

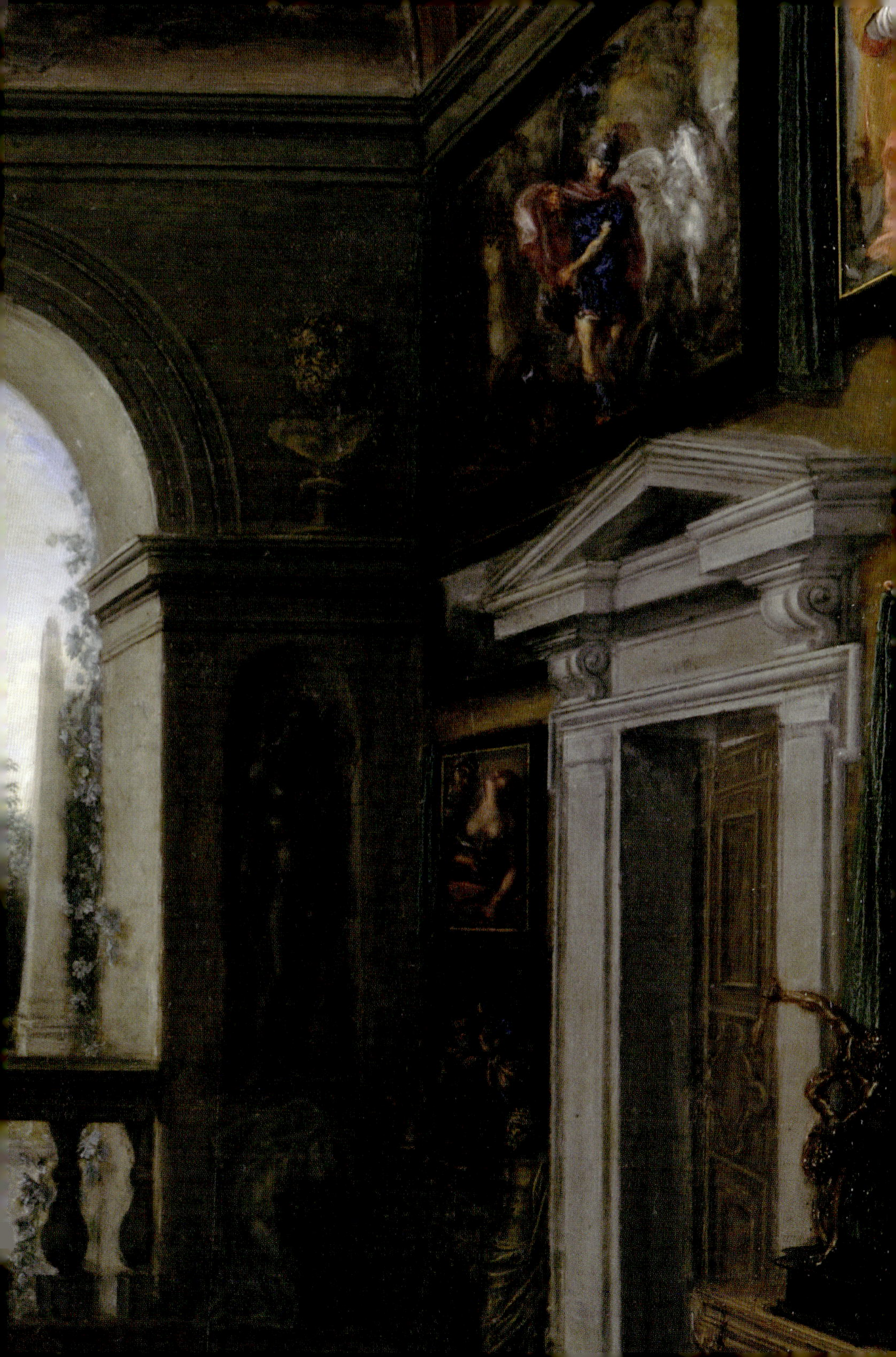

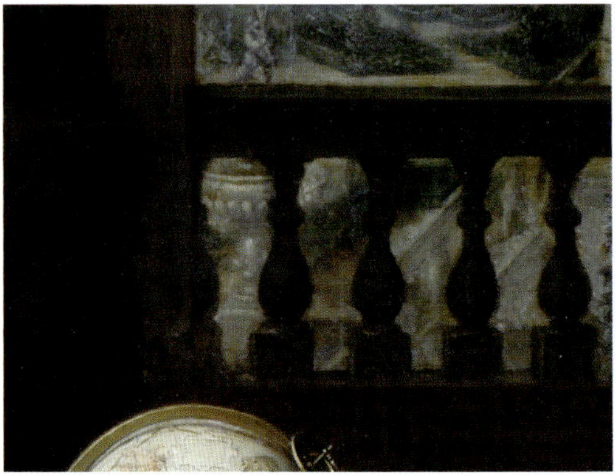

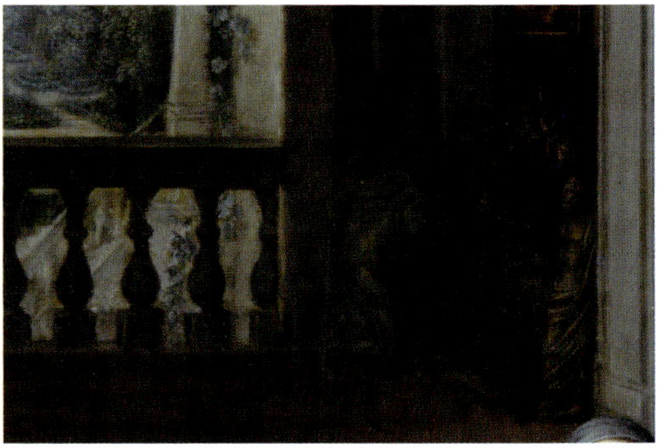

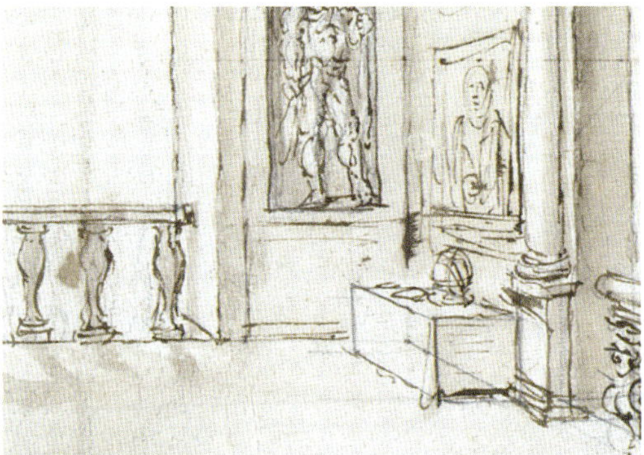

A Picture Gallery, Windsor Castle, RL 12983, detail.

changing temperature and changing atmospheric pressure. On account of the device Drebbel is often considered as one of the inventors of the thermometer.

GORMAN | Why is it near the window?

BRADBURNE | There are several possibilities for the placement of the *perpetuum mobile* near the window. In practical terms, as it worked in part due to temperature changes, it may have worked better where the sun could shine on it. The sun could also be used to reset the clockwork in the device. Symbolically, it may be related to the fountain in the garden seen outside the window, which seems indeed to be based on the famous 'singing fountain' outside the Belvedere in Prague, where Kepler and Brahe were known to have made observations. It may be placed in the shadows as it was not considered a real scientific instrument, but a wonder, a throwback to a questionable Rosicrucian past. Or it may just be a coincidence. The Windsor drawing also shows a spherical object in a similar position.

GORMAN | Was Drebbel also linked to Prague?

BRADBURNE | Rudolf's Prague was a known centre for all sorts of mystical and proto-scientific speculation, and Drebbel was invited to Prague by Rudolf II in 1610 to explain the workings of the *perpetuum mobile* to him in person. We know Drebbel stayed in Prague until shortly after Rudolf's death in 1612, after which he quite possibly made his way to the court of Rudolf's brother Albrecht in Antwerp, where this painting was probably made.

GORMAN | What makes you think Drebbel was connected to Antwerp?

BRADBURNE | Well, there is no documentary evidence that Drebbel worked in Antwerp. However, there are a number of unambiguous depictions of Drebbel's *perpetuum mobile* by the Antwerp artists Frans Francken II, Willem van Haecht, Hendrik Staben, Cornelis de Baellieur and Adriaen van Stalbemt, all painted in the 1620s and '30s. Although the cabinet painting genre was extremely popular at the time, it does not seem the *perpetuum mobile* was featured indiscriminately as a studio prop. And where there is a *perpetuum mobile*, it is likely that Drebbel was not far away.

GORMAN | How does it relate to the rest of the picture?

BRADBURNE | Drebbel's *perpetuum mobile* oddly only occurs in gallery interiors that have to do with either Albrecht and Isabella or Rubens. This is probably not just a coincidence. Albrecht is known to have had an interest in the *perpetuum mobile* as early as 1612, as Antonini presented him with the model he described to Galileo.

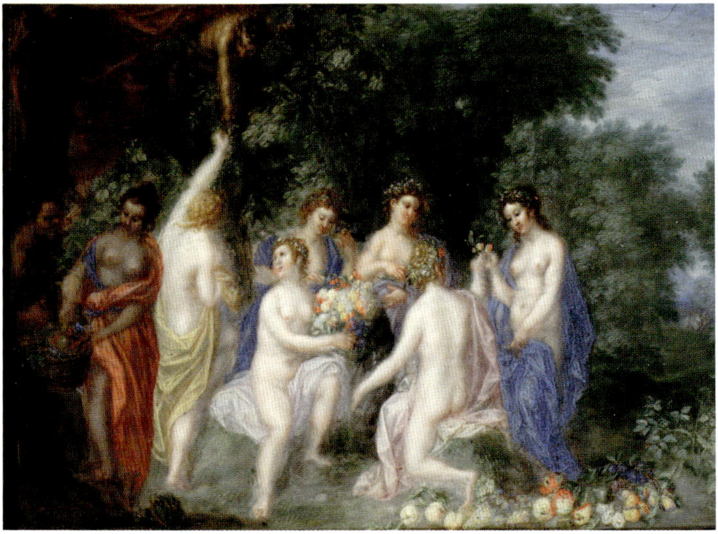

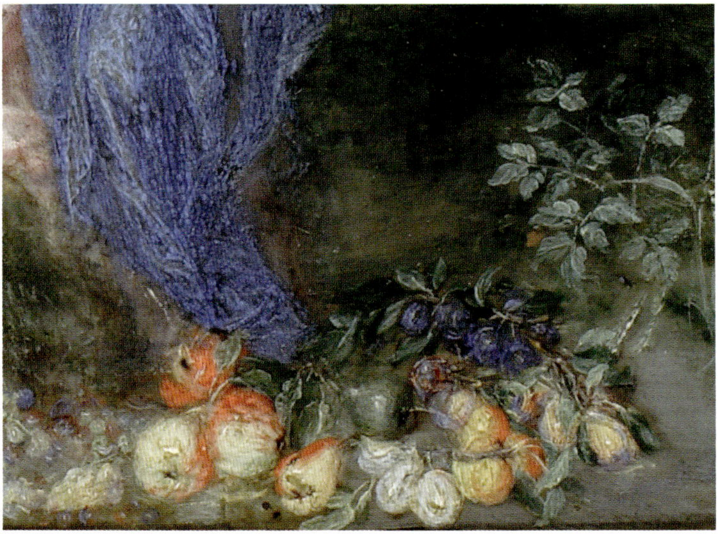

It is said that Drebbel made the Archduke a *perpetuum mobile* in 1615, for which Drebbel received a microscope in return, which while possible, seems unlikely, given that Drebbel himself is credited with the invention of the double convex microscope as early as 1609. The *perpetuum mobile* features in a painting of Archduchess Isabella alone in her study in 1627, in mourning after the death of Albrecht, so clearly the instrument was associated with the archducal couple. Rubens too was associated with the *perpetuum mobile*, as it was to Rubens that Peiresc turned when he wanted a copy made in 1624—we know from his letters that Rubens knew Drebbel personally.

GORMAN | Right, but what can we say about these paintings on the walls? Here we have nymphs and satyrs in the style of Hendrik van Balen, we also have this abundance which comes back to the theme of the cornucopia, the copiousness... You have all of these fruits, you have the mischievous satyrs—why is this picture here?

SMITH | It's a comment about nature and art as we have talked about, obviously. We have many manifestations of nature in each of the examples: the copiousness of nature, the generative properties of nature, the dangerous properties of nature, the ephemeral qualities in this wonderful still life here on the right side of the painting. This is an incredible still life.

GORMAN | Well, we have a number of landscapes, we have some still life, and we also have certain symmetries between left and right if one looks at the paintings. You have a flower painting on each side.

WESCHLER | You also have these two landscapes flanking that landscape.

GORMAN | Right.

WESCHLER | Once again that paradox. Two of them, in fact four of them are 'paintings', and one of them is 'nature'.

GORMAN | Then we have the *Triumph of Bacchus* here, with drunken Silenus and Bacchus being carried off, celebrating the fruits of the earth. Then we have the *Triumph of Neptune and Amphitrite*, parading the fruits of the sea. So there seems to be an interesting dialogue of juxtaposition going on.

WESCHLER | So how would you describe the relationship between the *Nymphs and Satyrs* on the left and the painting on the right?

GORMAN | OK, on the left we have pagan lasciviousness and bounty, and then on the right we have *Belshazzar's Feast* with all kinds of goodies and wonderful food, peacock pie and all sorts of riches, but he's suddenly seen the writing on the wall say-

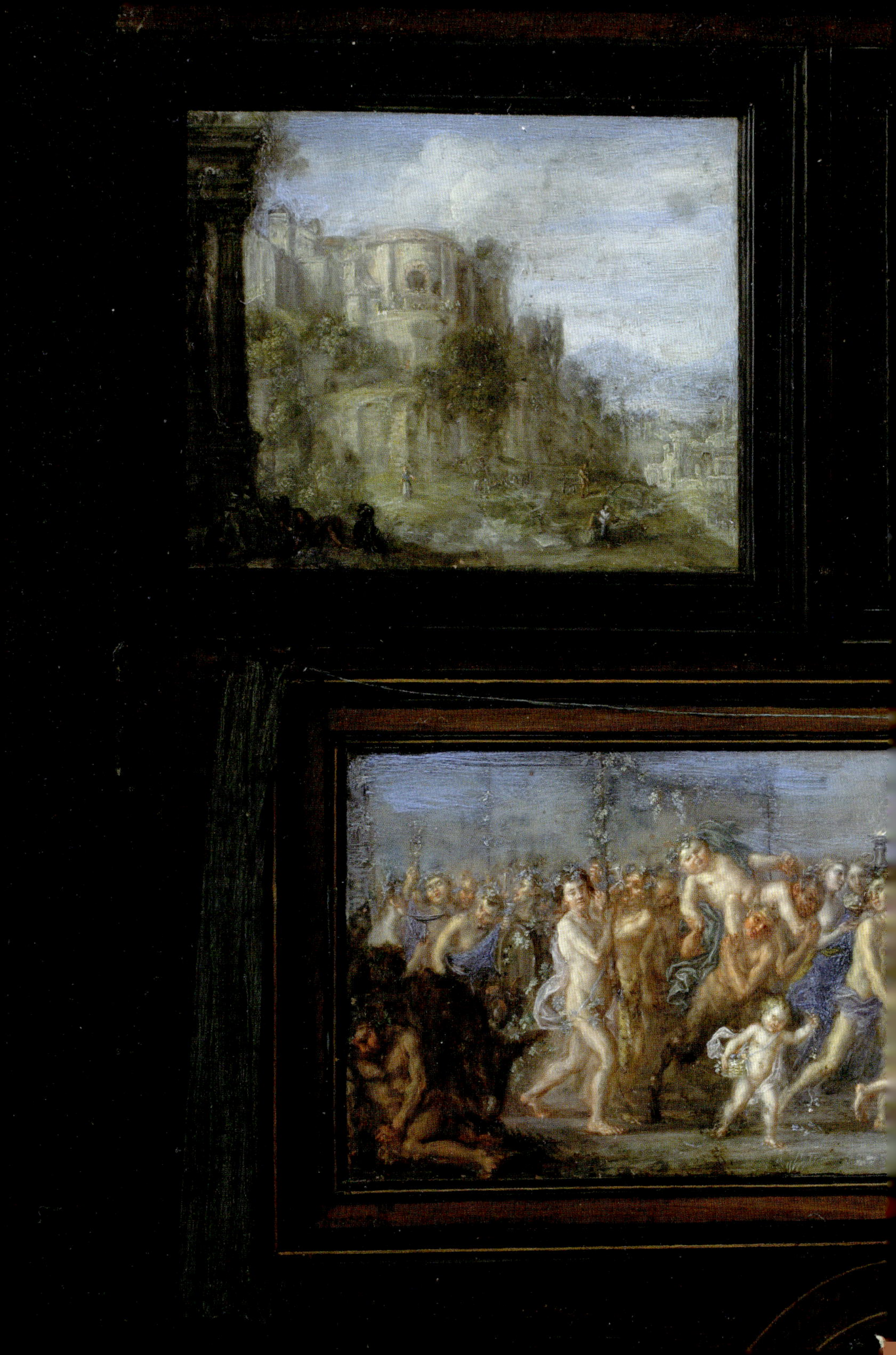

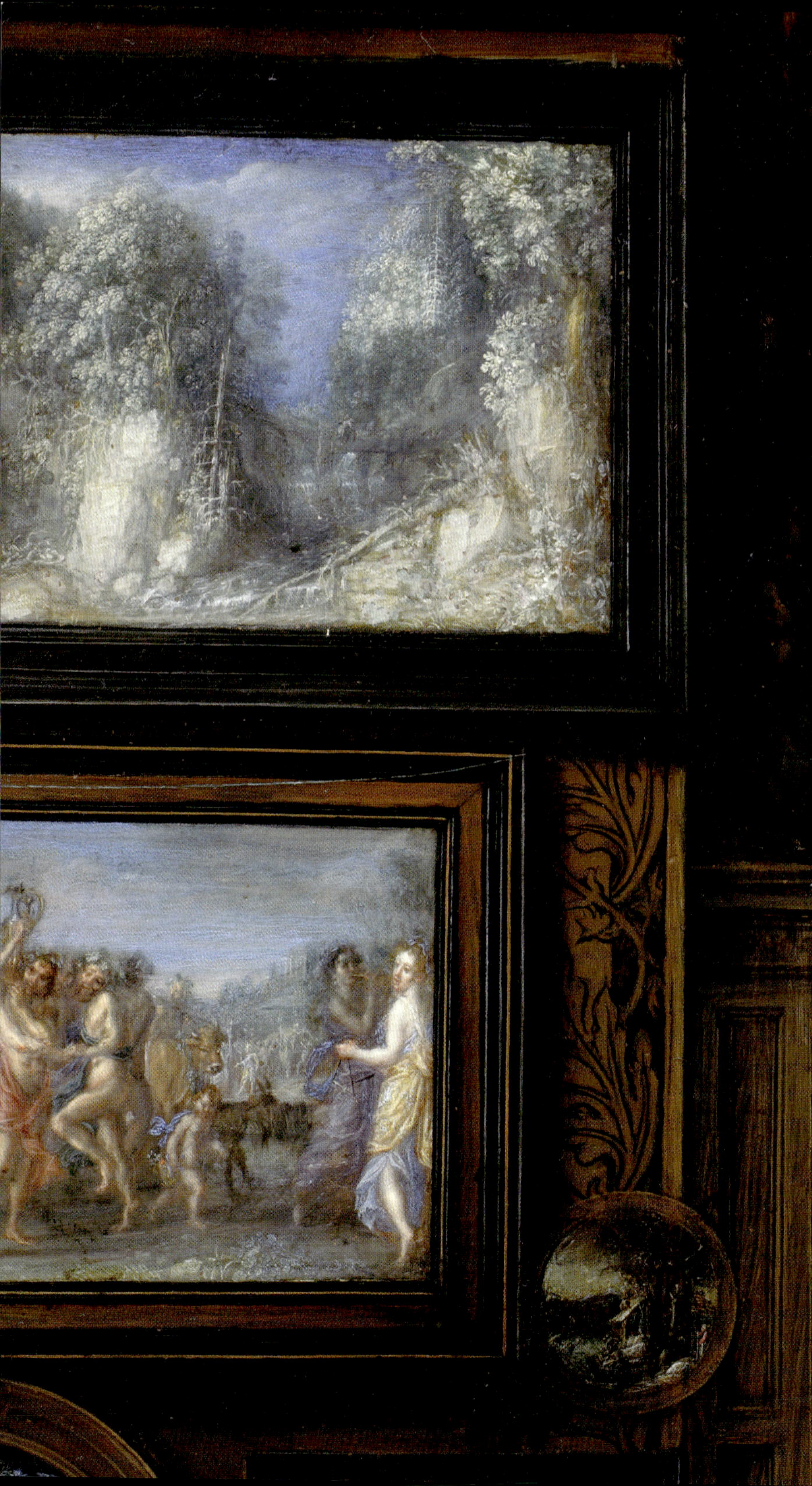

ing, "You have been weighed in the balance and found wanting". And so he's looking up, King Belshazzar, and he's actually looking out of the picture, he's looking up at the writing, which is very interesting, a stern admonition to excess. Are there any thoughts about whether these paintings actually existed? These could simply be meant as paintings in the style of various celebrated artists of the time.

WESCHLER | They could be the paintings he owns.

GORMAN | Are they real or imagined paintings?

MARR | It's a big question. The fact is that no paintings survive that seem to match the paintings in the picture, but that doesn't mean that they didn't once exist.

WESCHLER | The building could have burned down.

CORDOVER | Collections have been lost.

GORMAN | I believe there are compelling reasons to believe that they are not existing pictures.

CORDOVER | I agree with that.

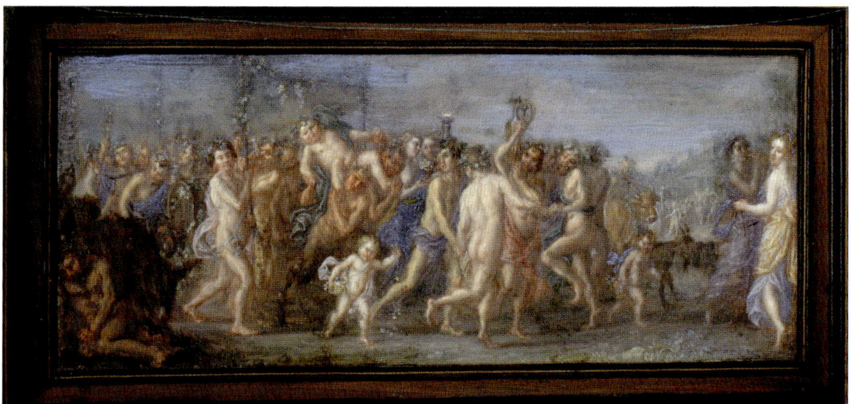

GORMAN | This is the drawing I discovered in the Royal Collection of Windsor Castle when I first began investigating this picture. It's a drawing which was attributed to Frans Francken II, but that's almost certainly an incorrect attribution. And it's a drawing that clearly has a lot to do with this painting. I mean, interestingly the space is not identical, but it's very similar. It doesn't have the same vaulted ceiling, it has a door on the left side rather than a window. It has the balustrade, the slightly recessed cupboard. But there are clearly very strong similarities. The scale is different; you can see that the people are slightly smaller relative to the room.

WESCHLER | Now your theory is that this came before that. A counter-theory could be that this is an actual space and this is what one guy did with it and this is what someone else did with it.

GORMAN | Well, why is it different if it's an actual space? Why do you change the ceiling? Why do you put a door in here?

CORDOVER | But the painting is an extraordinary intellectual effort and the

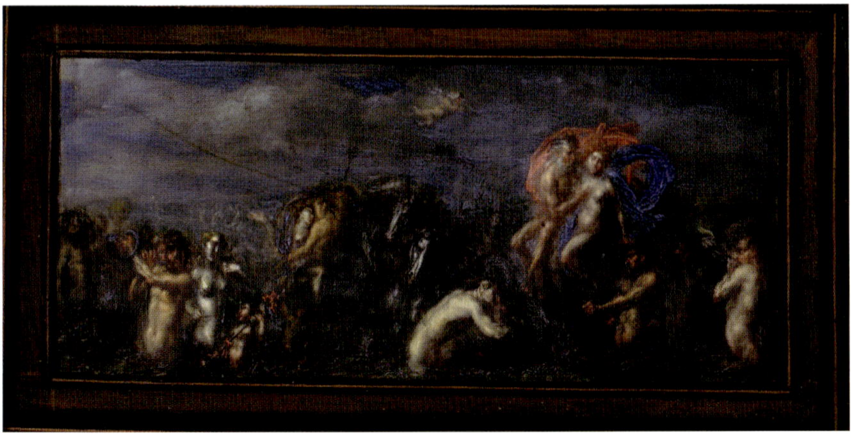

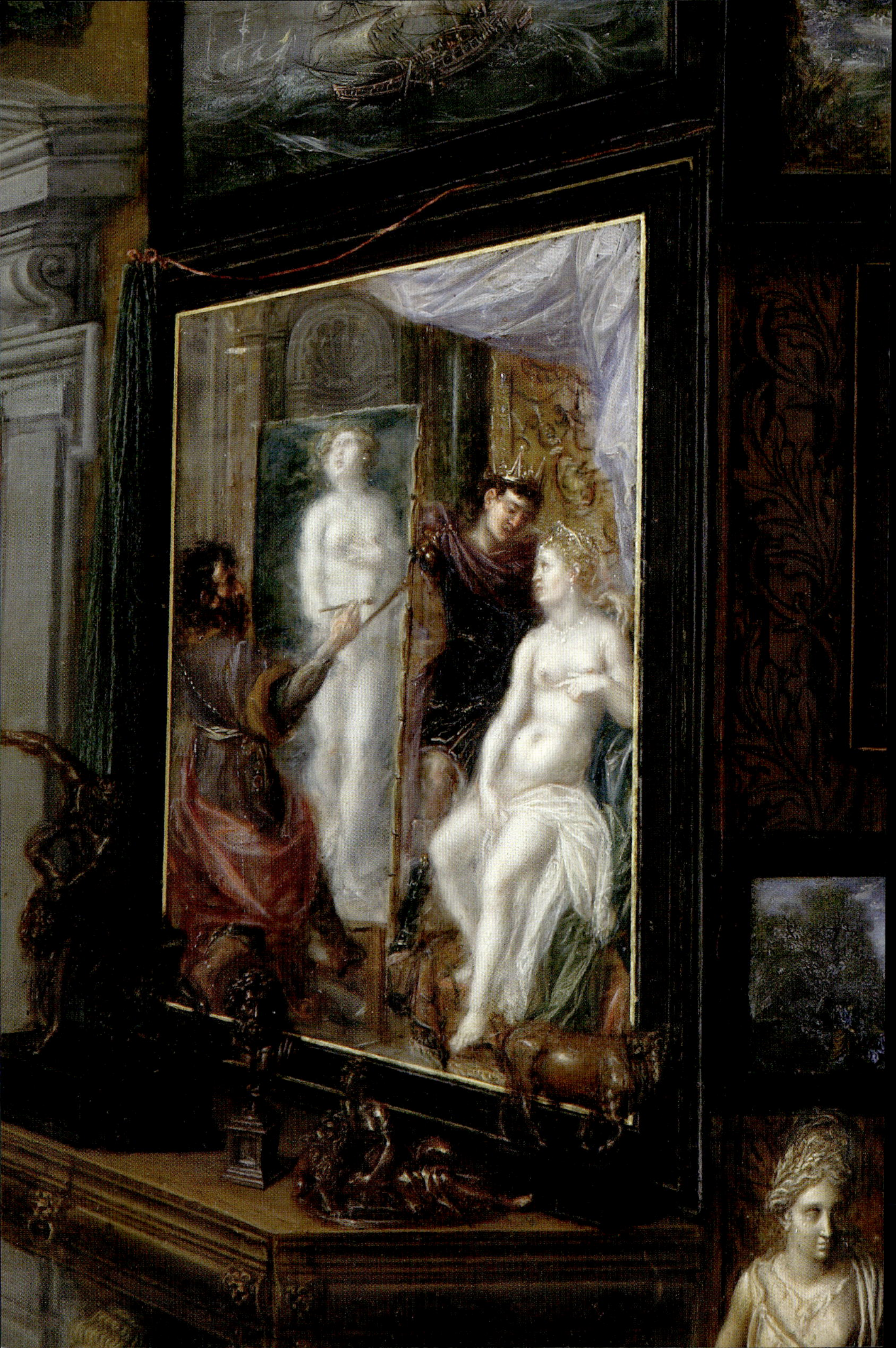

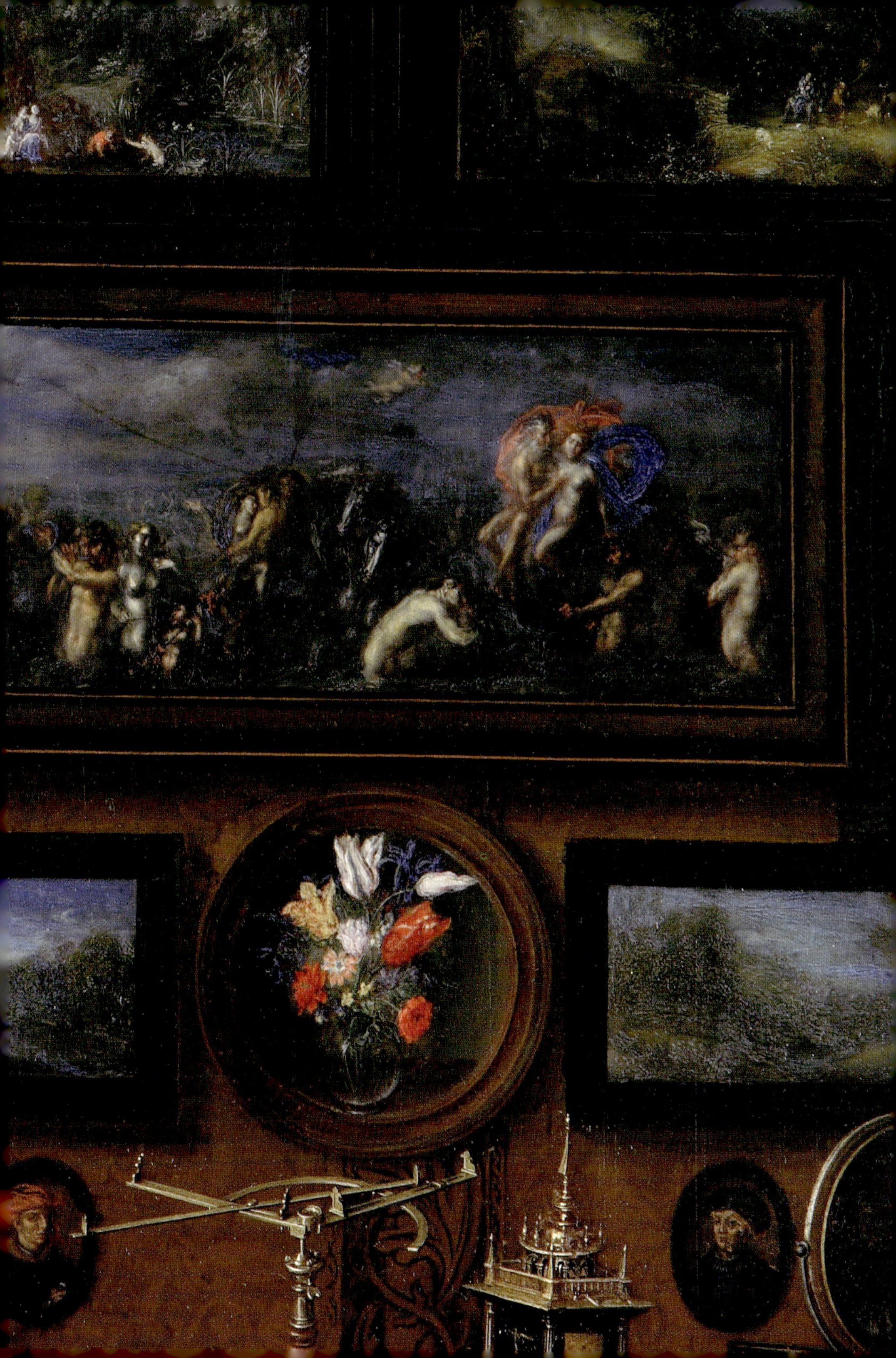

drawing is a much quicker study. Below the painting there is an underdrawing which mimics the perspective of the drawing…

GORMAN | If you look you can see that there are really quick sketches of some of the paintings on the wall and if you look at them really closely you can see that there is somebody reaching up to a tree and that there is somebody reaching down. But the arrangement, the composition is slightly different. It's not identical. So here you have somebody reaching up with the right arm and here you have somebody reaching up with the left arm. So it's clearly the same thought with slightly different composition. With *Belshazzar's Feast* you can see here there's somebody holding the peacock pie bringing it to someone sitting down and here the peacock pie is already on the table. So they're clearly not copies of existing paintings.

SMITH | That's fascinating.

GORMAN | They're compositional sketches that are then being transformed in the final versions. If you were going to copy a picture you wouldn't do it like that.

A Picture Gallery, WINDSOR CASTLE, RL 12983.

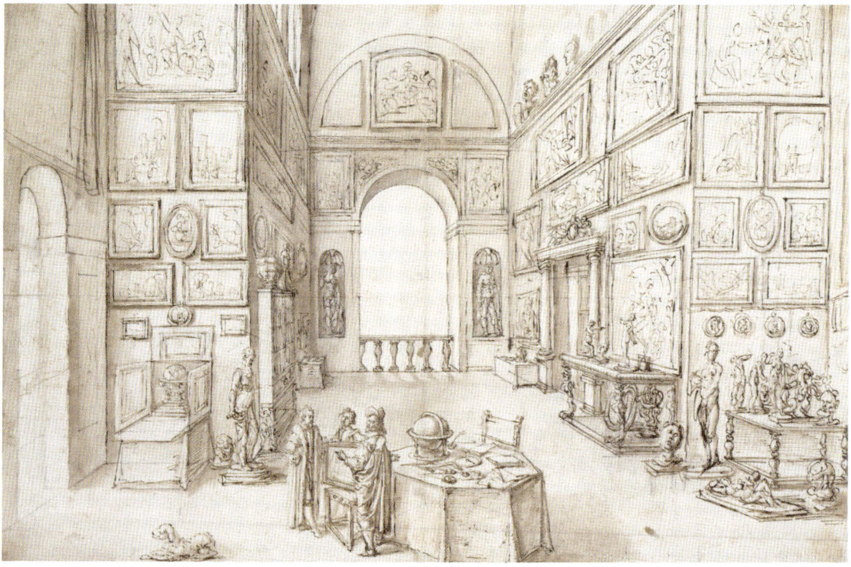

CORDOVER | The differences are interesting. You know, one of the pictures that is very prominently placed is the *Apelles and Campaspe*.

GORMAN | Yeah.

CORDOVER | And that doesn't appear to be the same picture.

WESCHLER | And here they have the Picasso. [*laughter*]

SMITH | See, I can imagine an artist bringing this drawing to Peter Linder and Peter Linder saying, well, let's think about the paintings. And then he says, but you know, I want my wall. I want my instruments and my globe and so on.

GORMAN | Well, I think we can come back to the causal link between the two because I think it's a really interesting question... But just in terms of the *Apelles and Campaspe*, that's a really interesting picture as well because you have Apelles, the most famous painter of classical antiquity, painting Campaspe, the mistress of Alexander the Great, and painting her so well that Alexander actually gave him the mistress as a reward for the quality of the painting.

SMITH | So where is its opposite?

GORMAN | I believe that its opposite is in the lunette above the balustrade looking out onto the garden. With Apelles we have the ultimate accolade of painting turned into flesh, and the ideal relationship between patron and artist, but up here we have an extraordinary thing—these people with donkeys' heads smashing works of arts and musical instruments and globes, in a vivid rendition of iconoclasm.

MARR | There's something rather odd about that lunette. The image of ass-headed men destroying art objects is a fairly common one in gallery interiors, showing ignorance destroying the arts and sciences. But here, and this is absolutely unique as far as I know, we don't only have ass-headed men—we have an ass-headed man in the centre, but the others are actual men.

GORMAN | It's clearly a reference to iconoclasm. There was the great iconoclasm of 1566 in Antwerp—long before this picture was made, but nonetheless a traumatic moment for Antwerp artists. But it's also interesting the way it's been painted. It's monochrome. It's almost as if it's been painted over. Almost as if the iconoclasts themselves have been effaced from the picture.

MARR | Yet we should also say that grisaille painting of that manner is standard decoration for rooms of this period. So it fits in entirely with this sort of architecture.

WESCHLER | The other thing you would say is not that they have been paint-

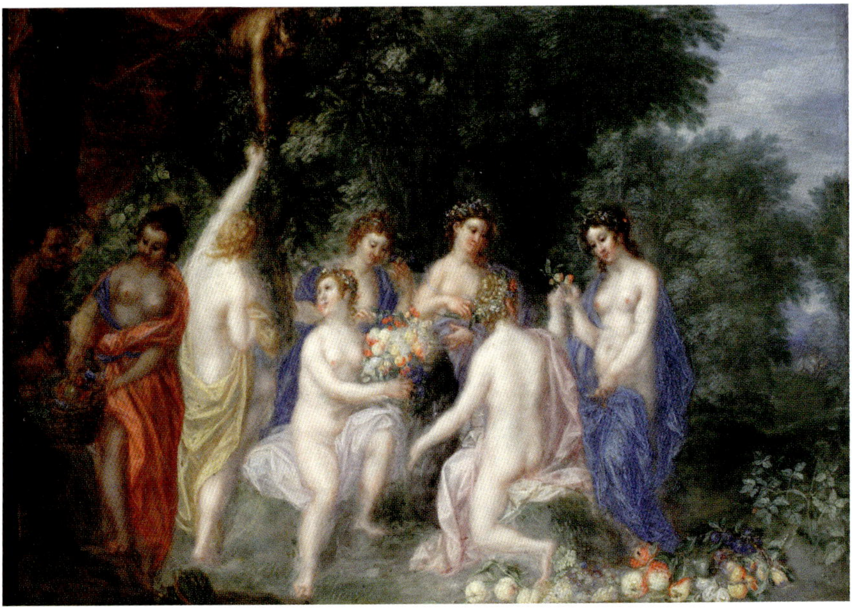

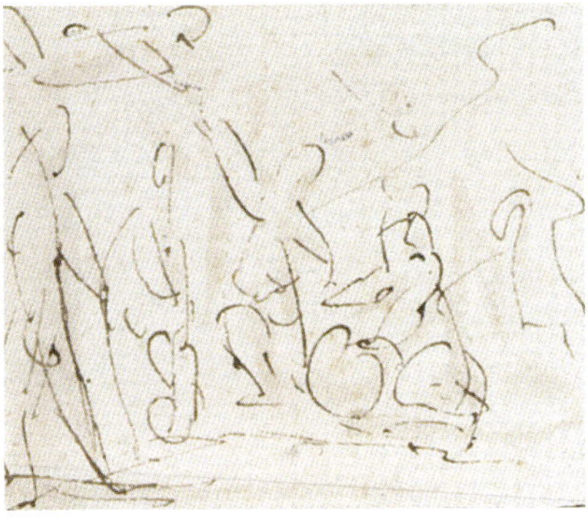

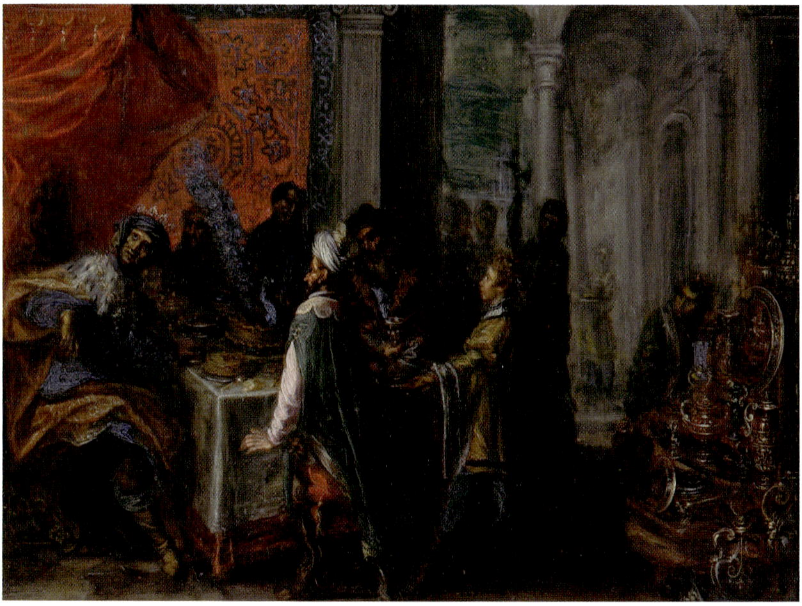

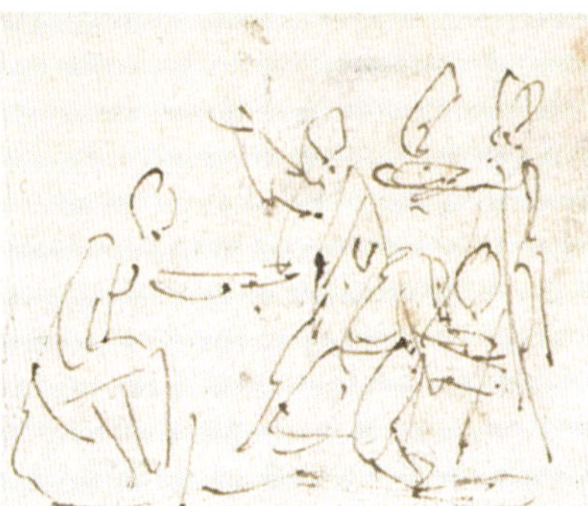

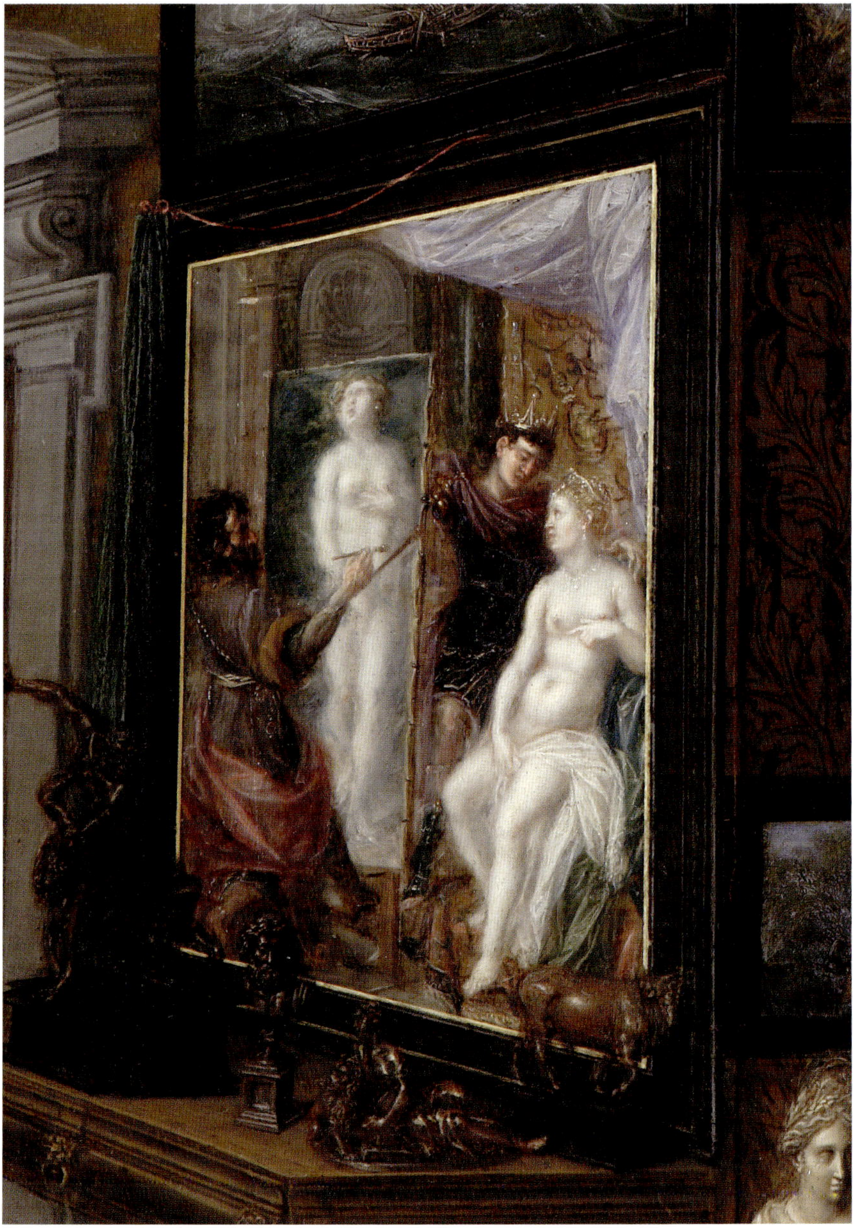

ed over, but they are in shadow. You've got two different worlds, and in one you have light overbrimming and in the other you have darkness and gloom.

GORMAN | And there's also the sense that this is on the boundary of the gallery. It's not really part of the collection. It's breaking the rules of the game—this is how not to consume art or how not to behave in a gallery. And the funny thing is these pictures don't exist independently of gallery paintings. They only exist in these kinds of pictures. Nobody has a picture of iconoclasm as a collection object.

MARR | Which is odd, isn't it. I've often wondered why. It's such an obvious subject.

WESCHLER | This is what it reminds me of—it's closest to Goya later on, that kind of violence, you'll get it later.

MARR | I have a theory that the Apelles painting is important because the artist and his patron were friends, as much as they were patron and client. And that's crucial to our understanding of how the *Linder Gallery* came about. It was a result not of

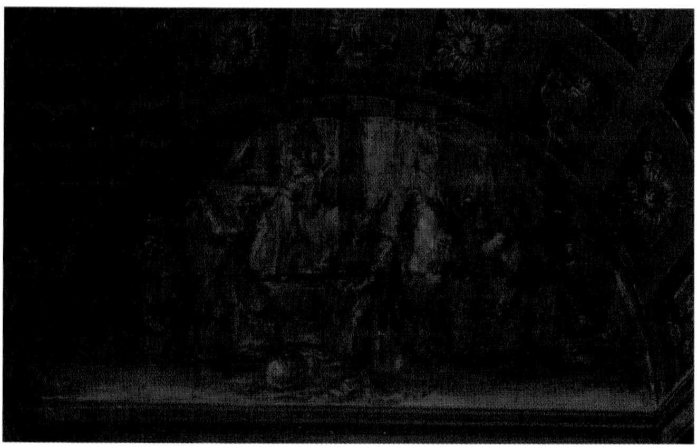

the sort of aggressive patronage that we're used to with some Renaissance figures, but of friendly patronage between equals. That was, in a sense, one the main points of the Apelles and Alexander story, that Apelles the humble painter was sufficiently equal with Alexander that he could be given his mistress.

GORMAN | If we look over here we have *Perseus with the Head of Medusa* with Pegasus in the background. So this is a story about optics, a story about mirrors, a story about using a shield to mirror Medusa's petrifying gaze. Then I think there may be another reference to optics here—in *Belshazzar's Feast* we have projection. We have the projection of these mysterious letters on the wall, and it's perhaps fanciful, but it seems to me that we have a figure here who could be manipulating a mirror to project the letters onto the wall as seen by Belshazzar.

WESCHLER | Imagine why I find that interesting!

GORMAN | Well, this is the technique used by Athanasius Kircher in the seventeenth century to project letters, as published in his *Great Art of Light and Shadow*.

WESCHLER | Of course.

GORMAN | Mirror writing. Rembrandt's painting of Belshazzar's feast also shows an example of this, a clear case of mirror writing with a finger.

SMITH | But it's not even a mirror. It's part of a banquet silver display.

GORMAN | It's a metal platter.

SMITH | Right, exactly.

GORMAN | Which could be used as a mirror.

SMITH | Oh yes, I know, but, it's wonderful because this is the kind of display that any noble banquet would have included. The display of gold and silver objects.

WESCHLER | One last thing. It's conspicuously in exactly the same relationship in that picture as is this down here.

GORMAN | Well, yes, and then we have a mirror in the painting.

WESCHLER | In the exact same location.

MARR | It does seem to me that it's a bit of a stretch because the platter in question is not smooth, it's chafed and has raised decorations. You would really struggle to get any kind of legible reflection.

WESCHLER | But look at his expression—that is a character if ever I saw one. He's up to something.

GORMAN | Let's look at the other paintings. We have the *Holy Family with St*

John the Baptist in a flower garland. And the detail of the flower garland is absolutely extraordinary. It's very hard to imagine that anybody but Jan Brueghel the Elder painted those flowers.

SMITH | Yeah, it seems like his signature. It just stands out of the painting so much, similar to the table.

GORMAN | This looks like a collaboration between Hendrick van Balen painting the figures and Jan Brueghel the Elder painting the flowers.

WESCHLER | But I would make the point about the lack of Christianity in this painting, or the paganism overwhelming the Christianity—what you look at here are the flowers. They just come so much to the fore compared to the Holy Family…

GORMAN | This was a kind of picture that was created at the behest of Cardinal Borromeo of Milan. It was his favourite kind of painting. Devotional painting. But it's a new kind of devotional painting, I suppose, the Virgin and Child in a gar-

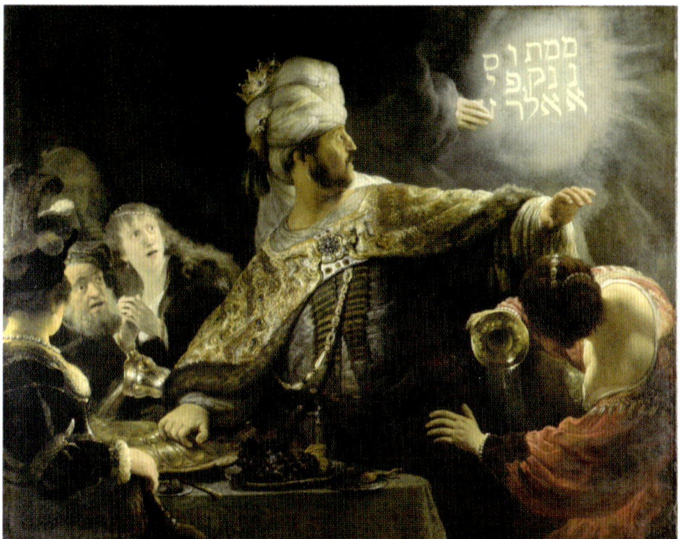

Rembrandt van Rijn (Leiden 1606–Amsterdam 1669), *Belshazzar's Feast*, 1635–6, oil on canvas, 167.6 x 209.2 cm. London, National Gallery, 6350.

land. It's outside of the ecclesiastical environment. It's a new kind of private collector's version of the devotional painting.

WESCHLER | In other words, I guess I'm just suggesting that we are on our way toward the secularization of Christian imagery.

GORMAN | Then we have these rather smaller pictures. It seems like we've got a portrait of Albrecht Dürer, what could be Jan van Eyck, and what could possibly be Lucas van Leyden. Why are these here?

SMITH | This is referring to the Northern tradition of painting. Not just a tradition of painting, but a way of seeing the world—a new way of seeing the world—and a new way of understanding the world and making knowledge about the world. And this was a recognized genealogy of painters from Carel van Mander's writings. Jan Brueghel the Elder is in that same genealogy. Any patron who wants to make reference not just to Rome—to antiquity—but also to the North is also going to have this genealogy of, well, painters/scientists. I want to say scientist, but it's totally anachronistic.

WESCHLER | OK, but look at this: two small paintings, and above, small landscape, small landscape, and above them flower, flower. That's why I want to make an argument that the floweriness of this is overwhelming the Christianity.

GORMAN | Well, I think there is definitely a secular–sacred dialogue going on. I think one can also see it in the landscapes, and if you look very closely at this landscape, it's actually Tobias and the Archangel. I can't quite make out the fish, but it's definitely Tobias and the Archangel. If you look very closely. Then you have secular landscapes on the other side.

WESCHLER | What have you been drinking? I'll have some of what he's having.

SMITH | There's one thing in what you say about the sacred–secular dialogue which is of course that this incredibly loving portrait of flowers became something that was bought in the Dutch Republic, where the art was not in churches, but in people's homes in great quantities. So you can say something about the use of paintings and of images in churches which was very much on people's minds at that time, I would think.

WESCHLER | But by the way, isn't it kind of striking how in a house that has shall we say—conservatively—thirty paintings on the wall, there's not a single one of a Crucifixion?

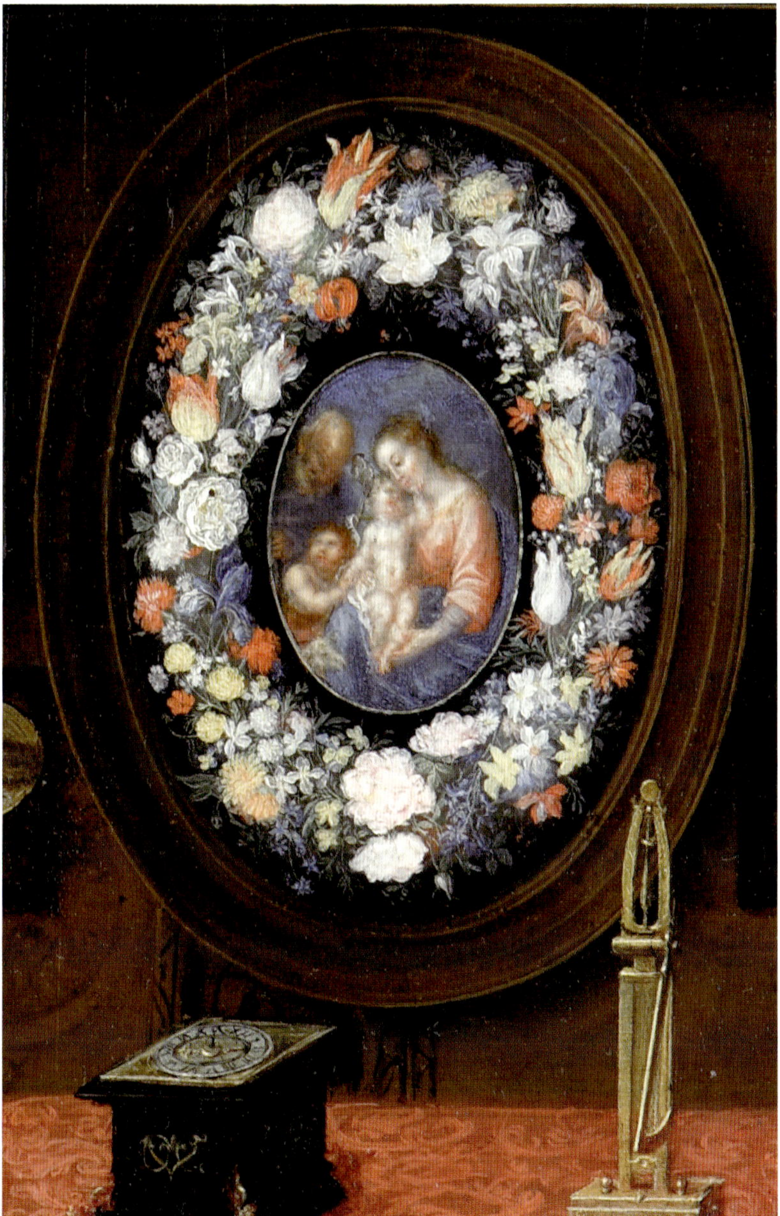

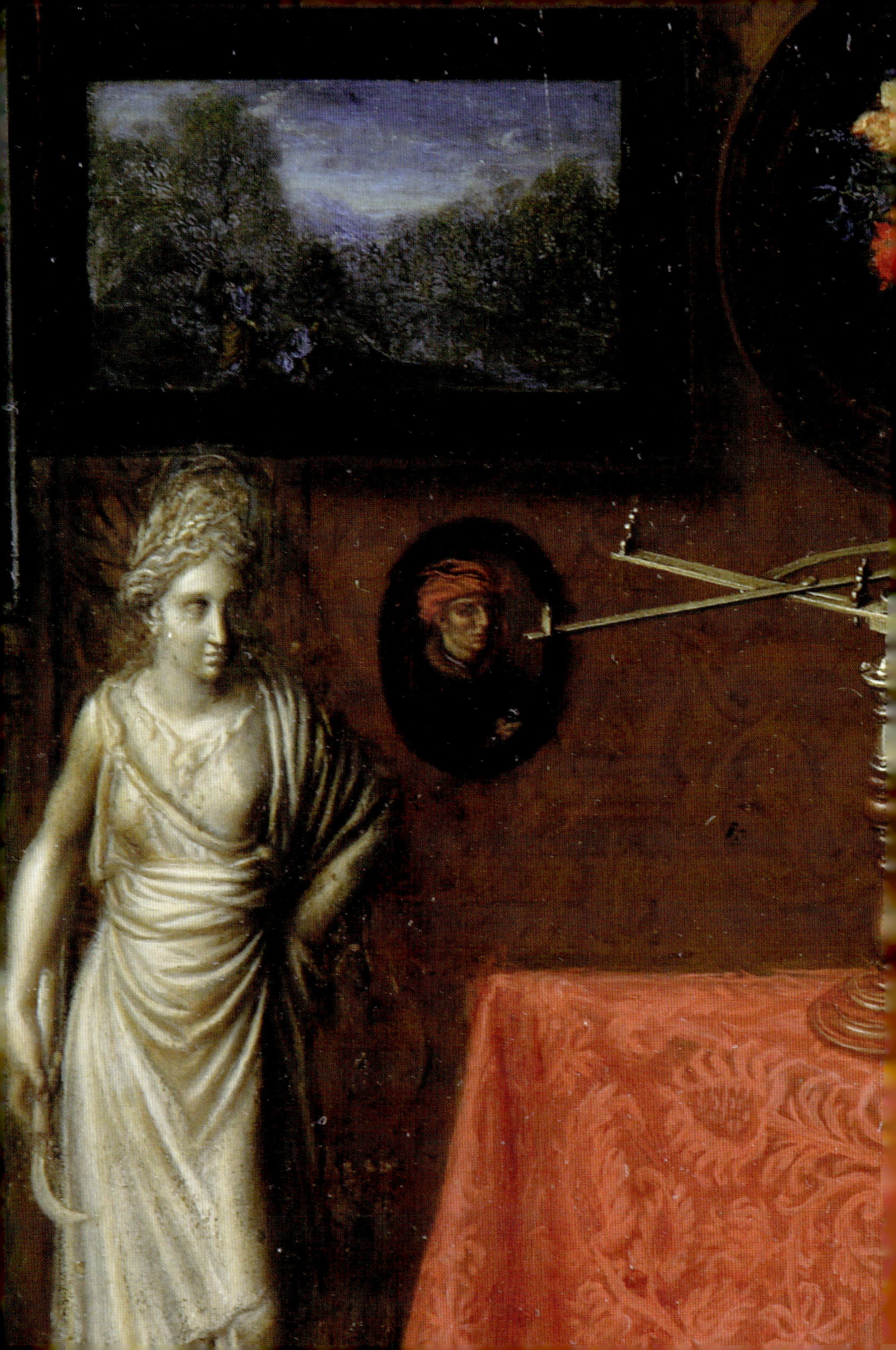

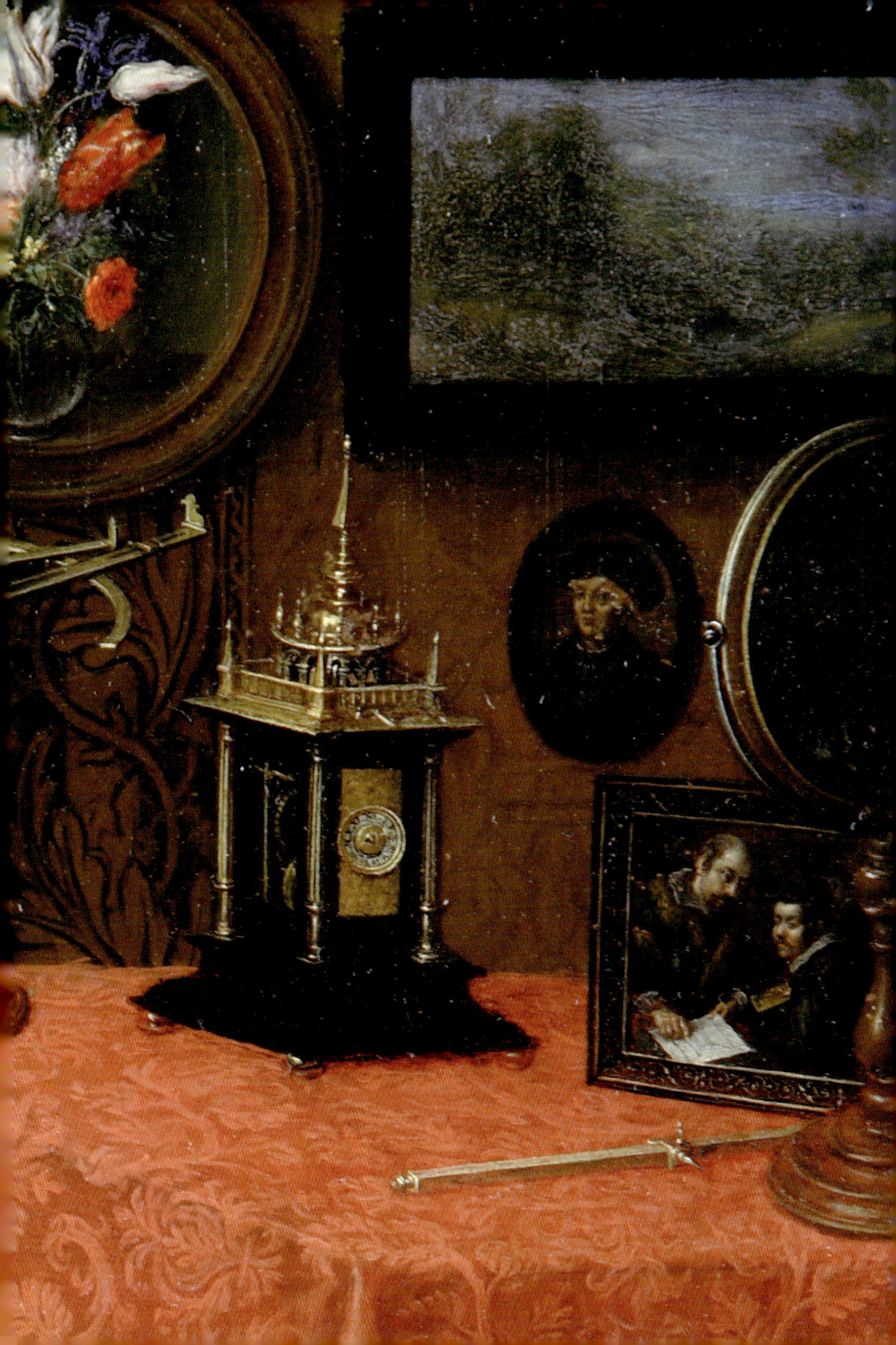

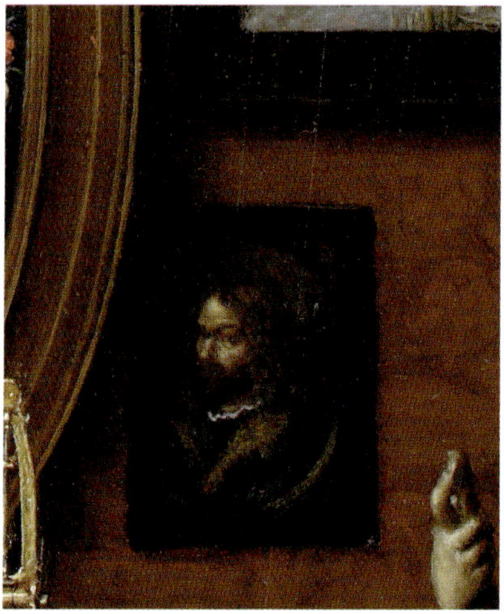

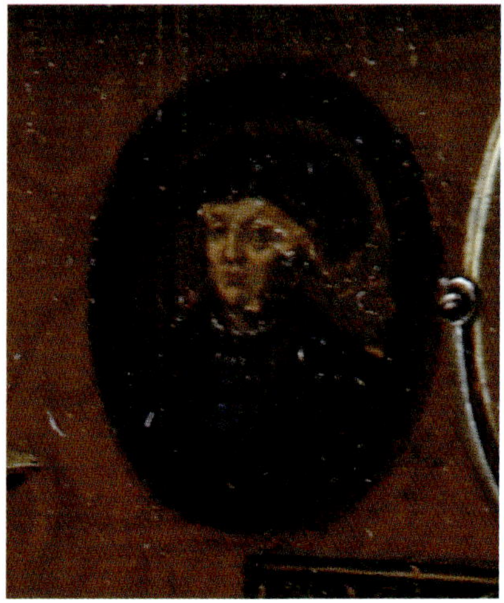

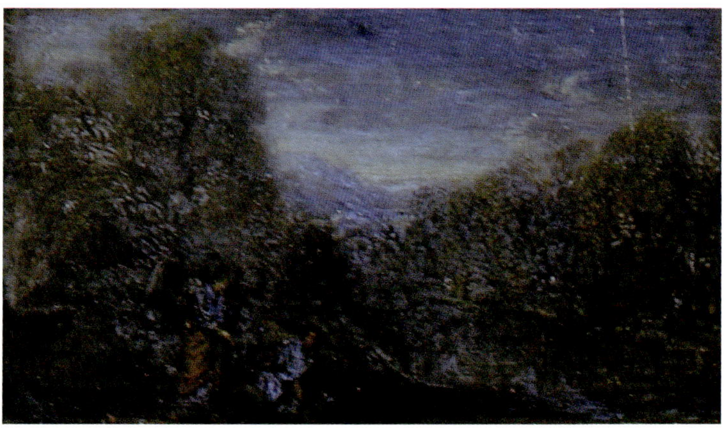

MARR | But the images we're seeing here are the sort of devotional images that we'd see in a bed chamber. It would be very common for you to have your Crucifixion or your penitent Mary Magdalene, the sort of images that we think of as being more fervent in that regard, in a separate space.

WESCHLER | I just keep coming back to the fact that Galileo is about to go on trial, this is in the very midst of the Counter-Reformation, all kinds of stuff is going on. That's all about to happen, and you've got all the tools here, and it's conspicuous that there's none of that here.

MARR | There is the issue of relevance.

CORDOVER | But there is, there is relevance. That's not fair, because he was having problems from 1615, 1616 on. And this was clearly done after that, even though it's prior to the 1633 trial.

MARR | I think no one expected—

WESCHLER | The Spanish Inquisition.

MARR | No one expected it to end the way it did. Most people thought it was a little spat which would end quietly.

GORMAN | What about the sculpture in the room? If we start on the left, we have possibly Pan. Then a shadowy figure of Apollo, and then we have Minerva, and possibly Amor. And then again over on this side there seems to be a symmetry. We have Bacchus and Ceres. There's a sense of the fruits of the earth balancing the fruits of the mind, perhaps. You have wisdom and love.

MARR | Sculptures are often used allegorically in gallery interiors. So I think we should be alert to the possibility that these, like the other objects, play some kind of meaning. I think that you're right to point to abundance and abundances of different kinds.

WESCHLER | As you've been doing this, the old joke about the turtles all the way down keeps on thrumming through my head. You know the story? The king can't sleep at night. He calls the wise man and he says, "I can't sleep. The Earth is falling through the heavens, what could possibly be keeping it up?" And the wise man tells him not to worry, the globe is on the back of a giant. But then the king can't sleep the next night: what's holding up the giant? He's on the back of an elephant. And so forth across successive nights—don't worry, the elephant's on the back of an eagle. The eagle's on the back of a walrus. The walrus is on the back of a turtle. And after that, don't worry, *it's turtles all the way down*. So, similarly, in terms of the 'what is this?', 'what is this?', 'what is this?', I want to make a critical point. There's a quality of et cetera here, of conspicuous et cetera. You get the idea that every one of these is saturated with meaning, blah, blah, blah, etc. And *that's* part of the meaning of the picture itself.

SMITH | Allusion and understanding the allusions is what connoisseurship is all about in this setting.

GORMAN | So what about the central octagonal table? On it we have this large Arsenius astrolabe, and then a celestial globe showing the constellations. This instrument appears to be a Galilean compass with an alidade, and then we have these lenses. You know, we were just looking at the picture with a magnifying glass and you have this array of lenses and magnifying glasses…

SMITH | You're supposed to pick them right up.

GORMAN | And with them you could examine these portrait medals, which are extremely precisely depicted. They're so accurately painted that every single one of them is an identifiable portrait medal. So we have Albrecht Dürer, Michelangelo Buonarroti. We have Girolamo Cardano.

WESCHLER | Who's almost falling off the table?

GORMAN | That's Michelangelo. He's very central.

MARR | Yeah. That's a nice idea of Michelangelo on his way off the table.

GORMAN | We have Bramante, Cardano and Alciati. And then one that's very difficult to decipher.

WESCHLER | So what are the books?

GORMAN | We have three books on this table which have titles that can just about be read. One of them is the *Rudolphine Tables* published by Kepler in 1627. One of them is the *Harmony of the World* also by Kepler, published in 1619. And then the uppermost book, which only a couple of weeks ago with the help of Paolo Galluzzi in Florence we discovered is actually John Napier's *Logarithms* of 1618, which were used very extensively by Kepler in doing the mathematics of the Rudolphine tables. So there's a very strong Kepler connection.

WESCHLER | But on top of everything is that weird drawing. What's that about?

GORMAN | It's actually an engraving of the martyrdom of St Catherine, by the Master MZ.

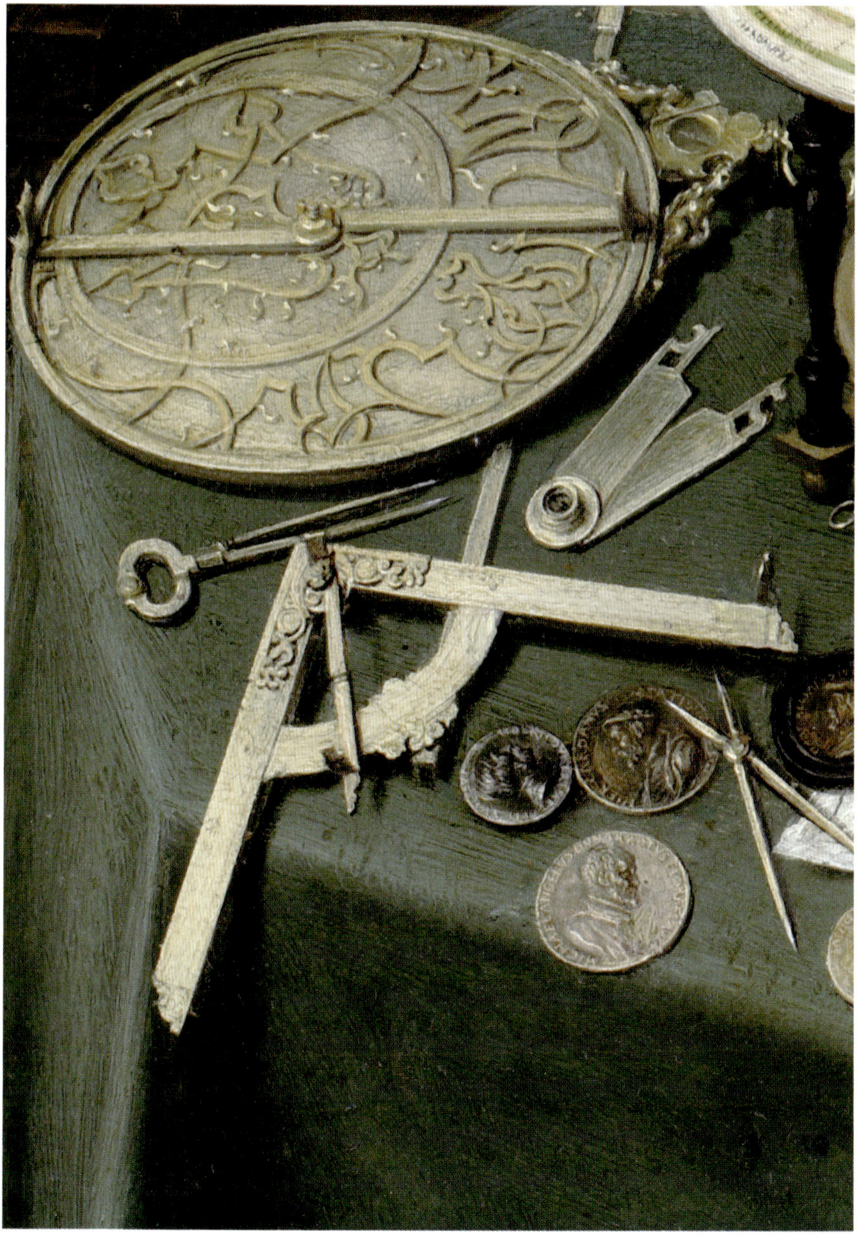

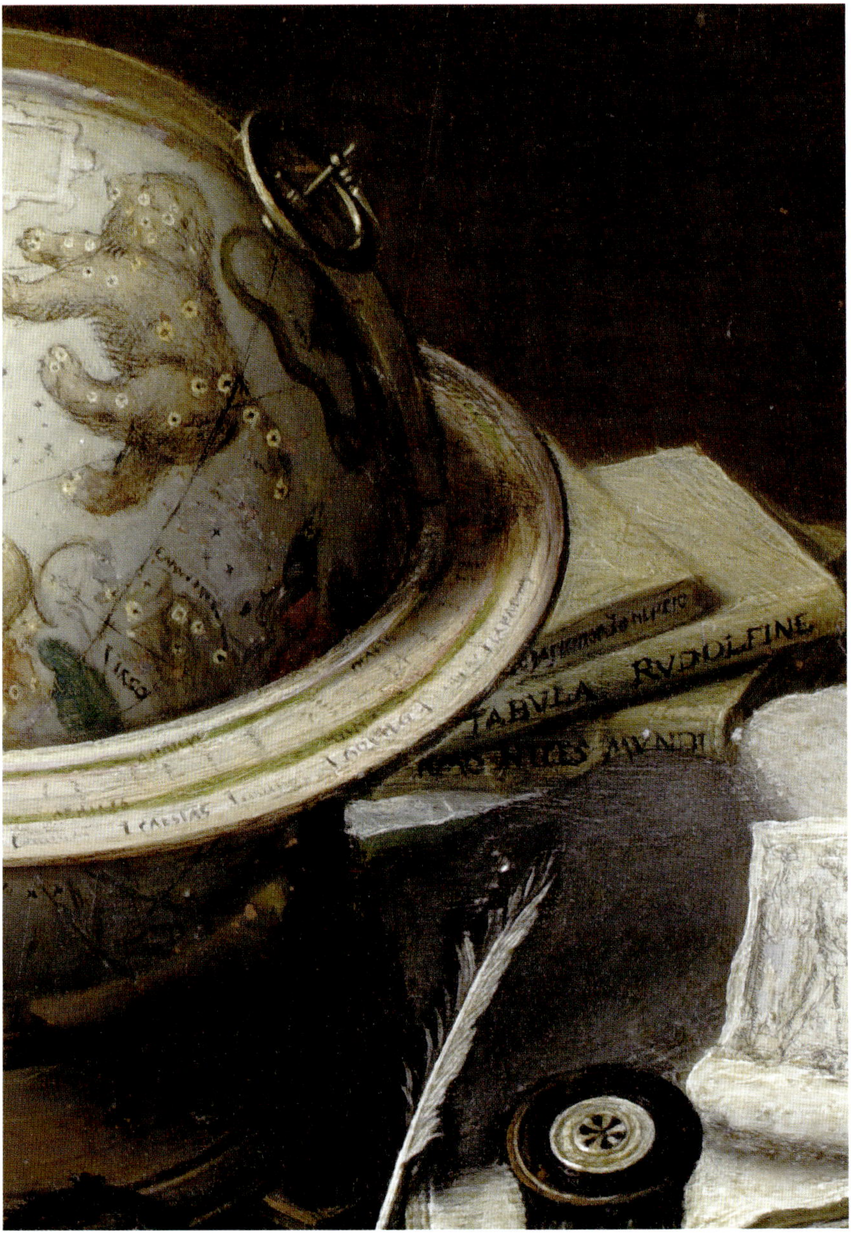
TABVLA RVDOLFINE
MVNDI

WESCHLER | So why is that there?

GORMAN | Well, St Catherine of Alexandria, St Catherine of the wheel, she was—they tried to kill her, to crush her on the wheel. But the wheel broke. So then they decapitated her and that worked.

CORDOVER | But before they did, she found herself at odds with the secular rulers, because she believed in Jesus Christ. And in order to make her recant her view, learned men were sent to her to persuade her that she had the wrong point of view. And she converted them to be believers, and that persistent point of view eventually led to her martyrdom.

WESCHLER | So why is this here?

CORDOVER | I think it's a warning. This picture is a warning to those who have unconventional views to beware.

WESCHLER | Or it's the opposite. That when people came to convince Galileo and so forth that he was wrong, he convinced them that they were wrong.

CORDOVER | Kepler's mother's name was Catherine and she was tried as a witch, so I think there may be some connection.

MARR | She was also the patron saint of certain mechanical artists.

GORMAN | And then we have the diagram of the world systems that we mentioned before. You have the Ptolemaic system on the left-hand side and the Copernican system with the Sun in the centre on the right. And underneath them you have the system of Tycho Brahe, with the planets going around the Sun and then the Sun itself going around the Earth. But then you have this cryptic Latin phrase: ALY ET ALIA VIDENT, 'Different people see it differently'. Or perhaps 'Others see it yet otherwise'—which could itself be a reference to Kepler. But it's an extraordinary line. It's a line which is saying that different opinions are possible about the universe.

WESCHLER | It's the place that a caption would be in a cartoon. It's the caption of the picture.

GORMAN | So what do you think about the phrase ALY ET ALIA VIDENT, Alex?

MARR | I completely agree with you that it's an exhortation. It points to the fact that there can be different interpretations. It doesn't necessarily condone different interpretations.

GORMAN | Right. Different people see it differently.

WESCHLER | And they better watch out.

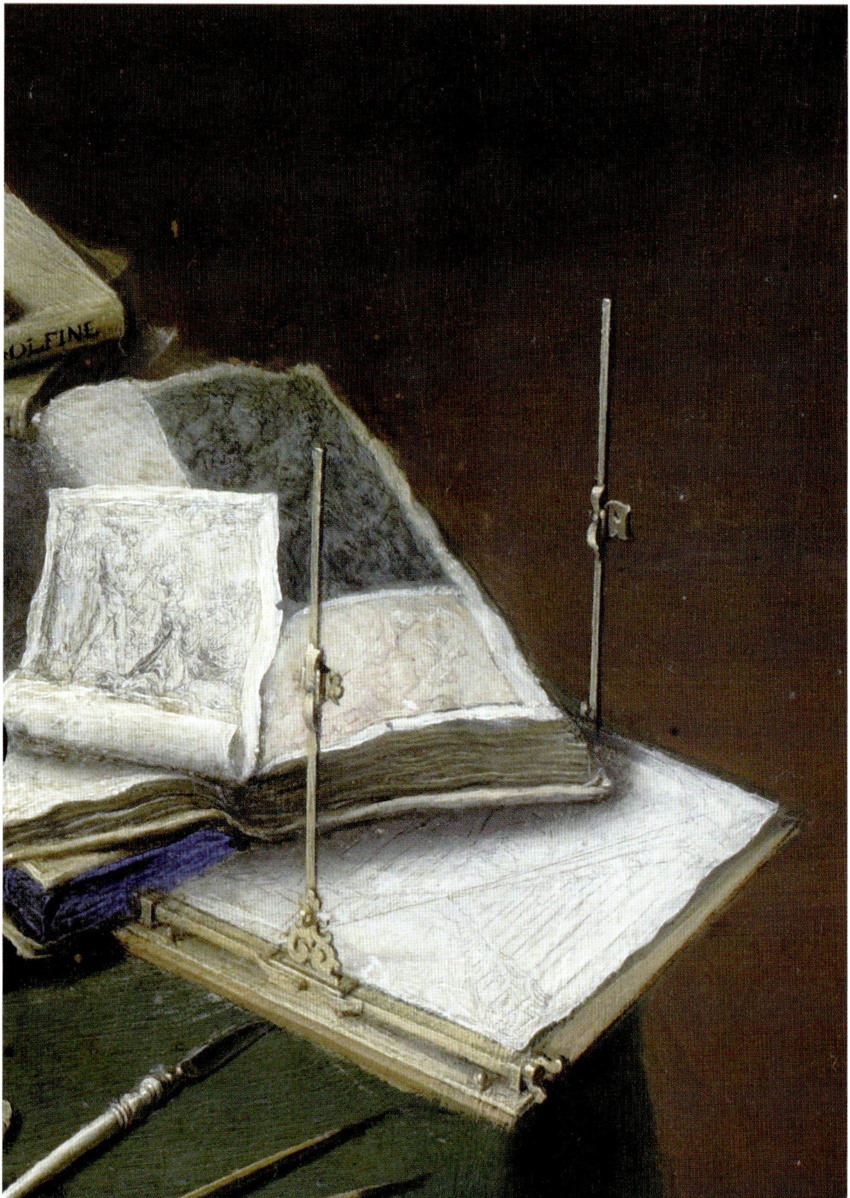
DOLFINE

SMITH | We really need to know the precise date and what's going on at this time. Because we're talking a lot about Galileo, but what really is going on?

GORMAN | Some of the books can help with the dates. The fact that the *Rudolphine Tables*, which were published in 1627, are there suggests that the picture was probably finished *post* 1627.

CORDOVER | But I would take issue with you about that, as Kepler was in the process of publishing *Tabulae Rudolphinae* for more than a decade before he finally got the money to do so, so it could very well have been well before 1627.

GORMAN | It really depends on what level of knowledge of Kepler one can expect on the part of the artist.

WESCHLER | One other point to make about dates is that we don't have CNN. So that something happens on a particular date doesn't mean that they hear about it in Antwerp.

GORMAN | Right. Partly hidden behind the three cosmic systems you have a horoscope, a geniture, showing the relationship between astrology and astronomy.

WESCHLER | Remembering that Newton had yet to come, and there is that fantastic phrase of John Maynard Keynes in his biography of Newton... Where John Maynard Keynes in the middle of the Depression, when we want him to be working on the Depression, takes six months off and goes and looks in Cambridge at all of Newton's notes, at the million pages—I am not making up that number—the *million* pages that preceded the *Principia Mathematica* and it is all over-the-top astrology and alchemy. And he says, "We have to realize Newton was not the first of the Moderns but the last of the Sumerians". This stuff bleeds, one into the other.

GORMAN | Kepler and Tycho Brahe were deeply involved in astrology.

SMITH | Yes, as was Cardano, so every astronomer that is being referred to there is making genitures.

GORMAN | The *Rudolphine Tables* were received as a great practical tool for astrologers. But what is going on with this drawing book over here?

MARR | Well, it is a common feature of gallery interiors. You often find open drawing books. Sometimes the images are identifiable, sometimes, as in this case, they are not.

WESCHLER | A drawing book means that it is a book that an artist was using to put drawings in?

MARR | No, it is like an album that a collector has into which he pastes drawings.

WESCHLER | It is almost as if the page on the left is blank and he is going to paste the picture of St Catherine onto that page.

MARR | It's curious. That's a nice idea.

GORMAN | What about the instrument that is under the drawing book—what do we know about that?

SMITH | This is an instrument that an artist would have used to construct a perspective drawing, and you can in fact see that there is a perspective drawing on it of a colonnade. This kind of instrument was definitely used by a Nuremberg instrument-maker and artist—an extremely well-known artist by the name of Wenzel Jamnitzer, a master goldsmith in Nuremberg.

GORMAN | Now I want to get on to something that is a critical feature of the picture. We have these two figures—an old man with a beard wearing a tabard and a young woman who is reclining on his lap. She has in her hands a pallet, paintbrushes, a maulstick and a sculptor's mallet, and there is also a book beside her. On her head she wears a laurel wreath and she has a sun pendant hanging around her neck. Who are these figures? What is their relationship?

WESCHLER | I want to make a strong case for this being a psychological portrait here, that it's not only that this means this and this equals this. And that all this connoisseurship is useless in terms of an existential relationship between an old man and a woman, you know, it only gets you so far.

CORDOVER | That's a very twenty-first-century interpretation.

WESCHLER | Well… Shakespeare would see it this way.

MARR | Shakespeare might have done, but Shakespeare was reasonably exceptional.

WESCHLER | Well, what I want to suggest is that Shakespeare is modern consciousness beginning to break out of all its contours and that's thirty years before this. And again keep in mind that this is an artist who is spending his whole day, week after week, working on this, different thoughts are occurring to him and so forth. There isn't just an X equals Y. At different moments they had different feelings about things.

CORDOVER | But I think the problem is not your assertion that there is a multiplicity of meanings. Because I think that all of us would agree that this has characteristics of representing the arts and drawing and the sciences, and while it has a contemporary real feel to it, the real feel is not a psychological reality…

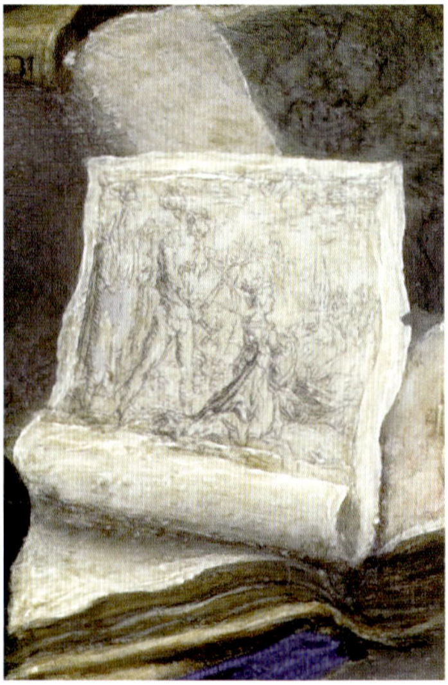

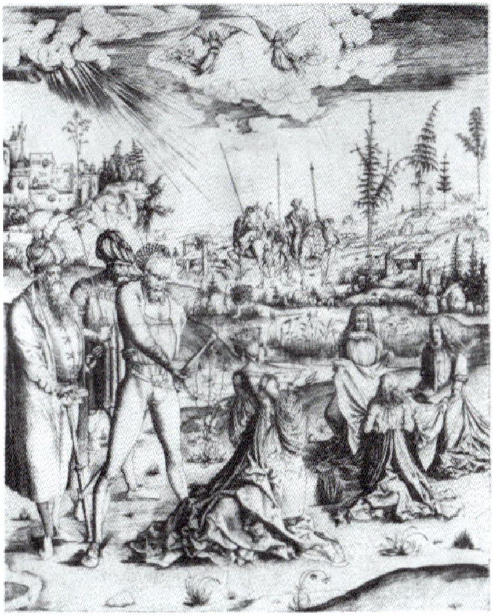

Master MZ
[M. Zasinger?],
Martyrdom of
St Catherine, *c.* 1500,
copper engraving,
309 x 254 mm.

Jost Amman (Zurich 1539–Nuremberg 1591)?, *Wenzel Jamnitzer and his Perspective Device*, 1568, engraving, from Wenzel Jamnitzer, *Perspectiva corporum regularium*, Nuremberg, 1568.

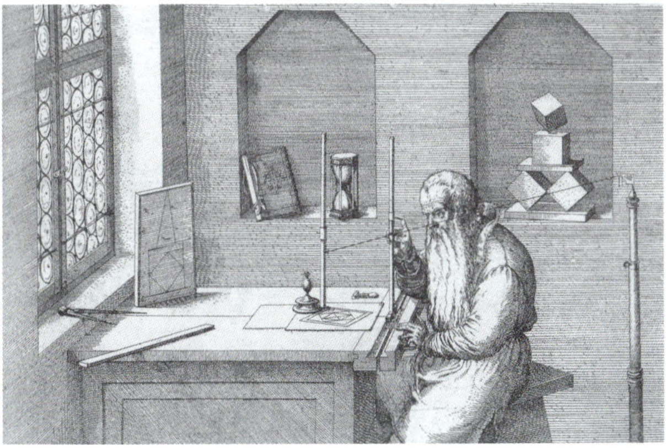

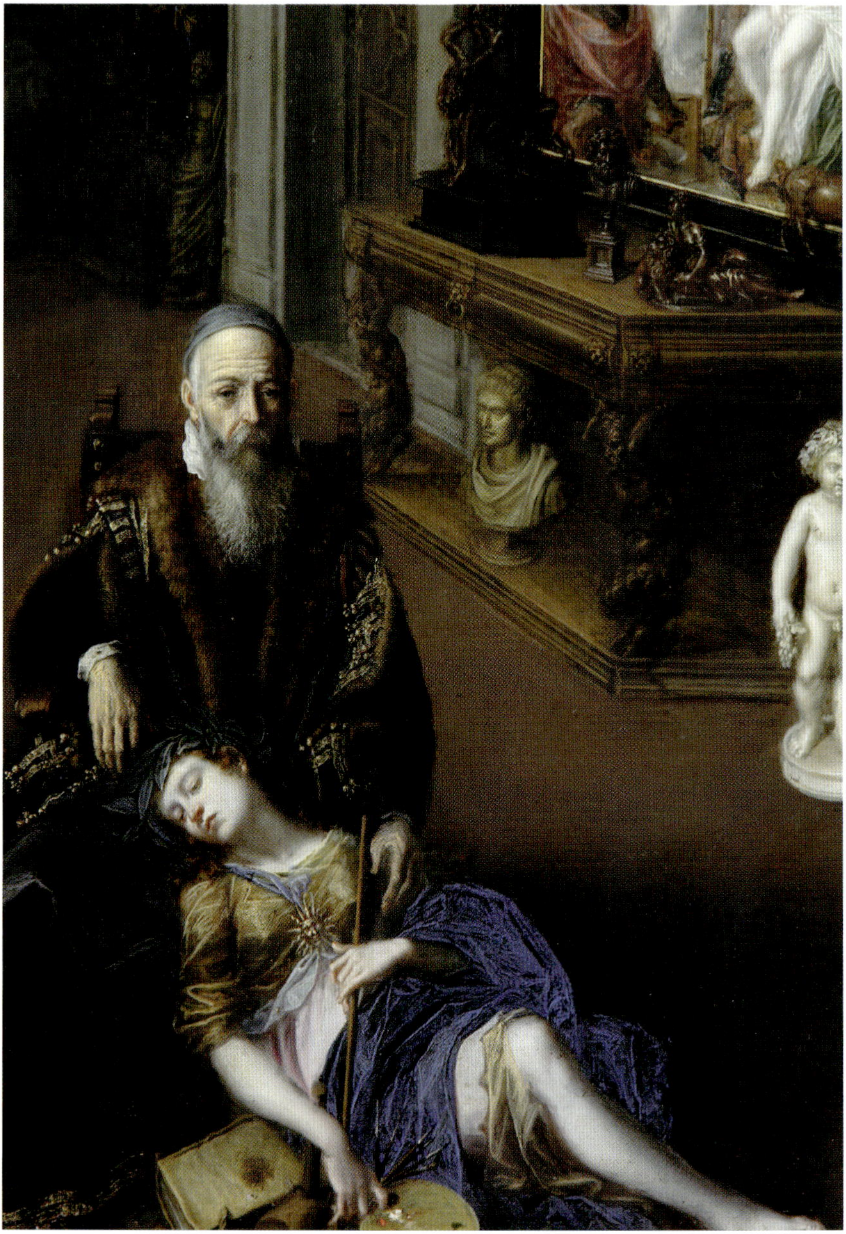

WESCHLER | Well, it's inhabited.

CORDOVER | But I don't think it's about our twenty-first-century sensibility about existence, about male and female.

WESCHLER | About mortality?

SMITH | Well, youth and age.

WESCHLER | What it is to be that old.

CORDOVER | And about—perhaps about one kind of knowledge versus a different kind of knowledge.

GORMAN | I think there is a sense in which the two figures in the foreground occupy a different space. They're bathed in a slightly different light. They're not really there in the sense that the figures in the Windsor drawing actually are there, even though they're smaller relative to the room.

SMITH | There is no question that the woman is Pictura. There is just no question.

GORMAN | She certainly is Pictura, or Painting, but maybe other things also. I mean, she's a sculptor, she has a mallet, what is the significance of the book? I'm not completely sure…

SMITH | Maybe she is image-making. Pictura, after all, is not just about painting. We think of 'to picture' as painting, but obviously Pictura means image-making.

GORMAN | Yes. And also the visual image projected on the retina in Kepler's description of the human eye. But the laurel wreath and the sun pendant, in Ripa's iconology, are attributes of Virtue. So as well as being Pictura she is also Virtue, and Virtue resting on the male figure representing Design, or Disegno. Do we think she is alive? I originally thought she was dead…

WESCHLER | Her eyes are alive.

GORMAN | She is a little pale. [*laughter*]

SMITH | She is pale because she is an allegory. But look at her lips, at her nose and her cheeks. If she were dead she would be greenish grey.

GORMAN | The male figure is painted in the manner of a portrait, as well as being allegorical. And if he is a portrait, there is a strong facial resemblance with Johannes Kepler. And Kepler is clearly strongly referenced in the octagonal table. And I think that there is an interesting thing about Kepler as the superstar of mathematics and astronomy in the seventeenth century, and the physical embodiment of geometry, astronomy and all of the arts that are so strongly represented in the painting.

There is a question as to whether he is intended to be a portrait or he may just be an allegorical figure. One suggestion is that he is Disegno—drawing and design.

WESCHLER | This is where I get off the train. Whether or not these have allegorical substrates, whether or not this is Kepler equals… This *is* design, this *is* drawing—no! This is a man, that is a woman. Look at them. There is an intense interior life going on in that picture. Even if it is Kepler, it is at a particular moment of his life, with a particular cast of interior life going on. There is tons of material life going on with her. They are not simply X equals Y…

GORMAN | That is true. They are not just ciphers.

WESCHLER | *That's* a cipher. Even that thing there is a cipher. But these people are alive. They are the two things that are living in the room. The world that I am describing is cracking out from this painting. The world that you're describing is embedded in the painting, it is a world of how things were, but there is a newer world coming, it is dawning, in the fact that her eyes are open and that she is thinking something!

MARR | This is awful liberal Whiggism! It's the march of progress towards a secular world…

WESCHLER | No! [*laughter*] I'm not saying it's the march of progress, but I'm saying that there is something different going on here. At this point historically.

MARR | Yes, but is this painting about that, or is it speaking to an old world order that should be preserved?

WESCHLER | In a way it is like we are having a conversation about *The Tempest*. Is Prospero just X, Y and Z, or is Prospero in a complicated relationship with Miranda, and it shifts from one moment to the next in the play, and at one point he is just a symbol of that and at another point he is something else entirely, and then he is something else again, and I love that all of that is happening.

MARR | Yeah. I think that there is a lot happening. It is almost like a theatre reference in the composition.

GORMAN | I think that there is a visual pun going on here. And it is a bad Flemish pun which says that painting and sculpture 'rest' on design, drawing and measurement. Not only painting, but also virtue, because design also means purpose.

MARR | Moral purpose.

GORMAN | Right. So, for example, if you kill somebody, the deed can be virtuous or vicious depending on the design or intention that you harbour.

SMITH | What other representations of design or Disegno in human form do we have?

GORMAN | Here is another one with Disegno depicted as an old man. This is another Flemish gallery painting which has been attributed to Stalbemt.

SMITH | Oh, interesting. What year, do we know?

GORMAN | Oh, probably early 1620s…

CORDOVER | Before, I would say.

GORMAN | And there is a similarity with the figures. Pictura has wings here and Disegno is more clearly allegorical in this picture because he has a flaming head.

WESCHLER | That painting seems purely allegorical to me. But *this* painting is something more.

GORMAN | I feel that this is portraiture. The man is a new kind of portraiture.

WESCHLER | But there is also interior life in these pictures.

SMITH | Look, it is quite a similar room. Even the ceiling looks the same…

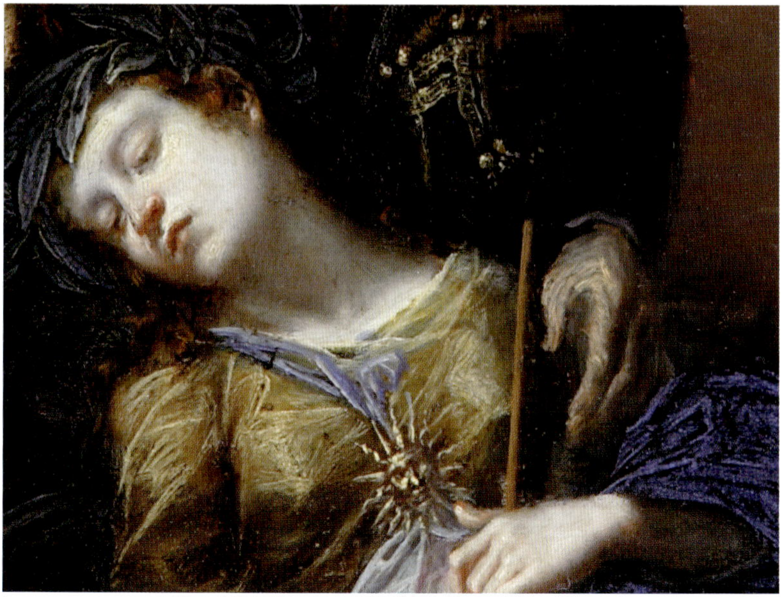

GORMAN | A little room here as well, with a vaulted ceiling and a tiny view out over the garden, with a little balustrade going down to the garden as well. So there are some quite interesting similarities. Some of these are real paintings in this picture. And here you've got your globe. But this is much more traditional—you've got your monkey and flowers and fruit and copiousness in the gallery.

MARR | And all of the painters that are painting in this genre—and there are only about six of them—are related to each other or knew each other or worked in each other's studios.

GORMAN | This is a portrait of Kepler in the frontispiece of the *Rudolphine Tables*. Kepler, wearing a tabard, sitting with a similar cap.

SMITH | I have to say—I'm sorry, Alex, but it looks very similar.

MARR | I personally don't think that this figure looks sufficiently like Kepler in the portraits. And there are lots of bearded old men who wore skullcaps.

GORMAN | Well, we have three books connected to Kepler on this central table.

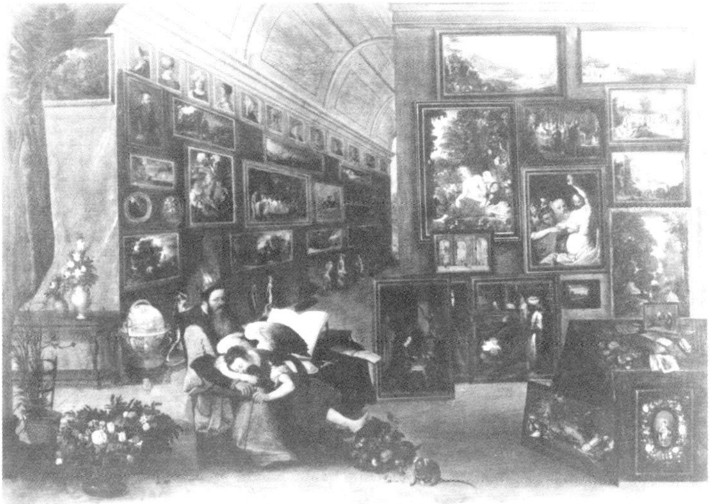

ADRIAEN VAN STALBEMT (ANTWERP 1580–1662), *The Interior of a Picture Gallery with Allegorical Figures*, OIL ON WOOD, 72.9 X 104.1 CM. CHRISTIE'S LONDON, 13 DECEMBER 1974.

If you look at the whole frontispiece of the *Rudolphine Tables*, you have the Tychonic system at the top here and then Kepler hidden at the base.

MARR | I think that there is no doubt that there is a Keplerian influence in this painting, but we should also point out that Michael John and I don't see eye to eye about the lettering of the books. There is the issue that the lettering on the spines of the books could have been added possibly at a later date. To my eyes, the lettering of the books and the spines is sufficiently different to the care and attention of lettering elsewhere in the picture, particularly cosmic systems, that even if it is clearer and it was intentional, it suggests that this isn't quite as important as other elements of the painting. Because if it is of supreme importance, you will take greater care over it. The fact that so many parts of this picture are painted with such deftness, such exactitude, says to me: if this is all about Kepler, why, when it comes to the key thing that identifies Kepler, are you sloppy?

GORMAN | I'm not saying it's all about Kepler. But if you wanted a person to

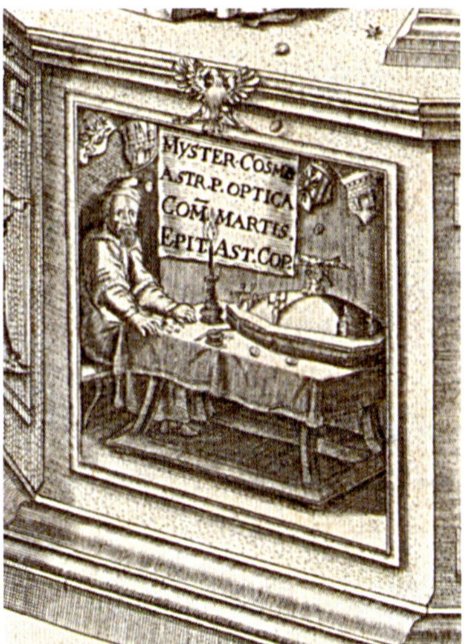

Georg Celer, *Temple of Urania*, frontispiece of Johannes Kepler's *Tabulae Rudolphinae*, Ulm, Jonas Saur, 1627. Florence, Biblioteca Nazionale Centrale di Firenze, Magl. 5.1.118.

embody mathematics, astrology, astronomy—all of the things that are so important to this picture—in the 1620s, who would it be? The imperial mathematician Kepler. He was the Einstein of the period, there is no question about it.

SMITH | Well, especially in the North, I would say.

WESCHLER | What do you make of Harold Bloom's comment, with reference to this picture in a sense, that before Shakespeare there were no interior monologues—that people didn't *have* them?

MARR | I would say that things can exist before you have a speech act to define them.

WESCHLER | But do you think that Shakespeare is evincing a different way of being in the world?

MARR | I think that Shakespeare is doing something rather different. I think that this one unusual example of a genre is not in any way in the same league and doesn't point to the same broad humanistic concerns that characterize Shakespeare.

GORMAN | I disagree with you, Alex. I mean, I think that there is something extraordinary going on here, where you have something which is a positioning of the art of painting in relation to the philosophical and cosmic discussions of the time, with this extraordinary human figure in the centre, and very different from the way allegorical figures are done in other pictures, the ones we looked at. It's powerful…

MARR | But Rubens uses portraits of friends as personification, so it's not uncommon.

WESCHLER | My point, though, is that if I were going to continue the interpretation I began with at the beginning of all this, that this is at least partially all about surfeit and too-muchness, if you look at that face, there is a sense of—I don't want to project too much, but I don't want to leave it out—there is a sense of having lived a life of too-muchness, and at the end of it all there is a sense of quiet—

GORMAN | Melancholy, perhaps.

WESCHLER | Melancholy, a kind of hollowness, a kind of giving that has famished the craving…

CORDOVER | I think we are reading a lot into a centimetre and a half.

WESCHLER | Well, that's my story and I'm sticking to it.

GORMAN | Let's go on now to one picture that we haven't discussed yet, which is this tiny little double portrait on the lower right-hand table showing two men—

one of them is pointing at a drawing, and the other man is looking at the drawing, and in his left hand he's holding a palette and in his right hand he appears to be holding a paintbrush. If one looks really closely, the drawing itself appears to be a perspective scheme for the whole room. It seems that we are looking at a portrait of a patron and the painter himself, painting the picture from a picture within the picture—this is an extraordinarily self-referential act. So there you see the perspective of the room itself. Who are these people? How can we identify them? Who is this patron and who is this painter? Alex, do you want to comment?

MARR | Sure. Michael John and I found out who these people were through a remarkable series of rapid exchanges

GORMAN | That's right.

MARR | Feverishly over one night, when I discovered a letter in an archive I was working in which was written to a mathematician and architect called Muzio Oddi, an Italian, by one of his pupils, a man called Giovan Battista Caravaggio. In this letter, written in 1629, he talks about visiting the studio, the study, of their mutual friend, the Nuremberg merchant Peter Linder. Linder was then living in Milan, as was Caravaggio, and Caravaggio describes a painting that he saw in Linder's cabinet which was hung with many pictures and beautiful instruments, and it contained within it a portrait medal of Muzio Oddi himself. And Caravaggio says, 'Oh, I understand from Linder you were responsible for devising the composition of this picture.'

GORMAN | I had written a piece about the painting and had flung a draft up on the Internet, with all the information I had been able to piece together. Then one night I got an e-mail from Alex saying, 'I found a letter which seems to describe this picture, but I am not sure if it is the same picture because it doesn't mention the figures.' I said, 'That sounds very interesting. Tell me about the letter.' So he said, 'It is from a man called Caravaggio and he is writing to the architect and mathematician Muzio Oddi, and he mentions that he's seen the house of Linder and he describes the picture with all the instruments and so on.' And then suddenly it dawned on me that I had been looking at the coat of arms in the left-hand window and it appeared to be the arms of the Linder family, so immediately, when he mentioned Linder, it was clear that this was the painting. So we had a rapid flurry of e-mails and discovered that the man on the left-hand side of the double portrait appeared to be a German merchant called Peter Linder, the patron.

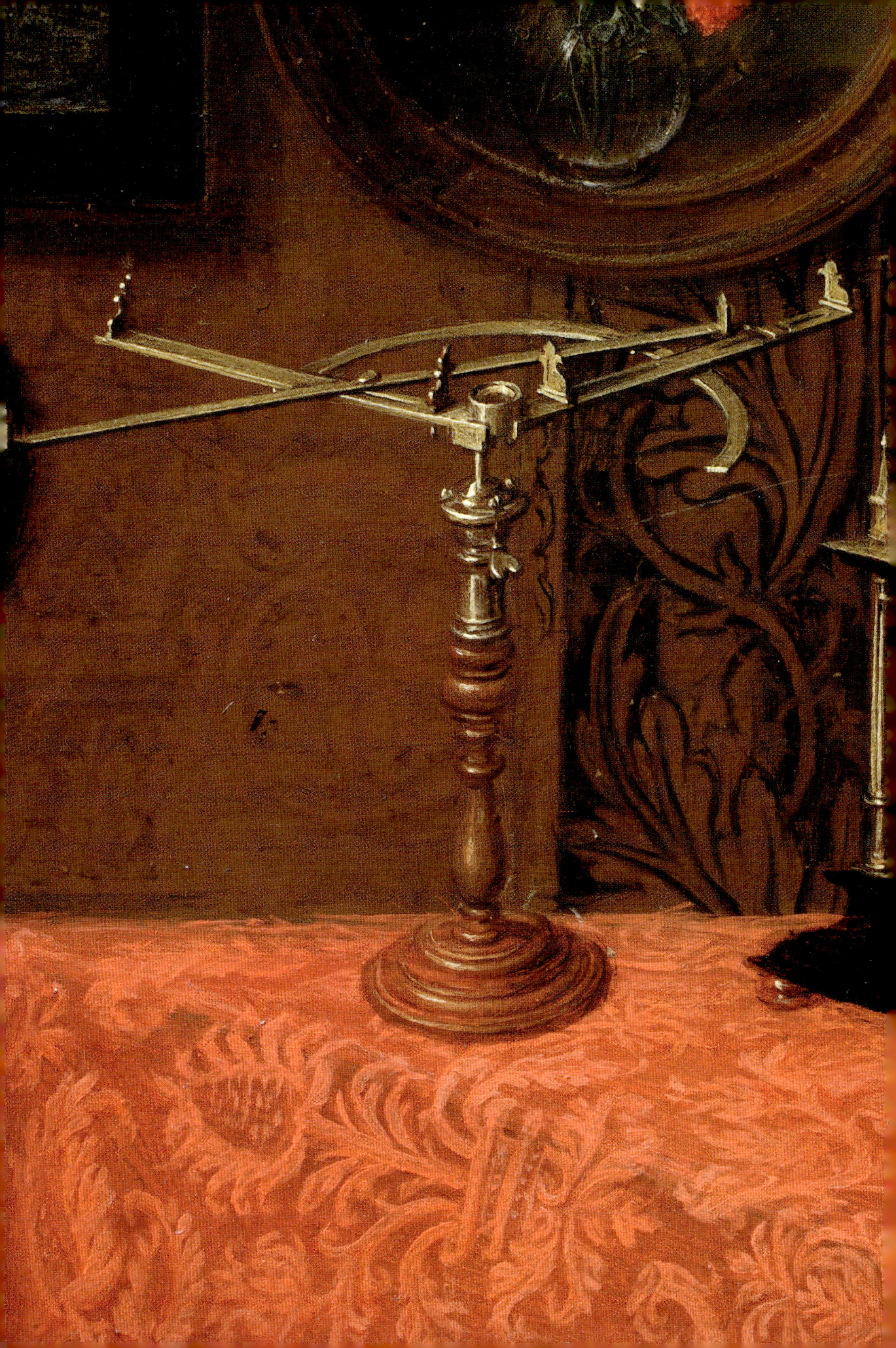

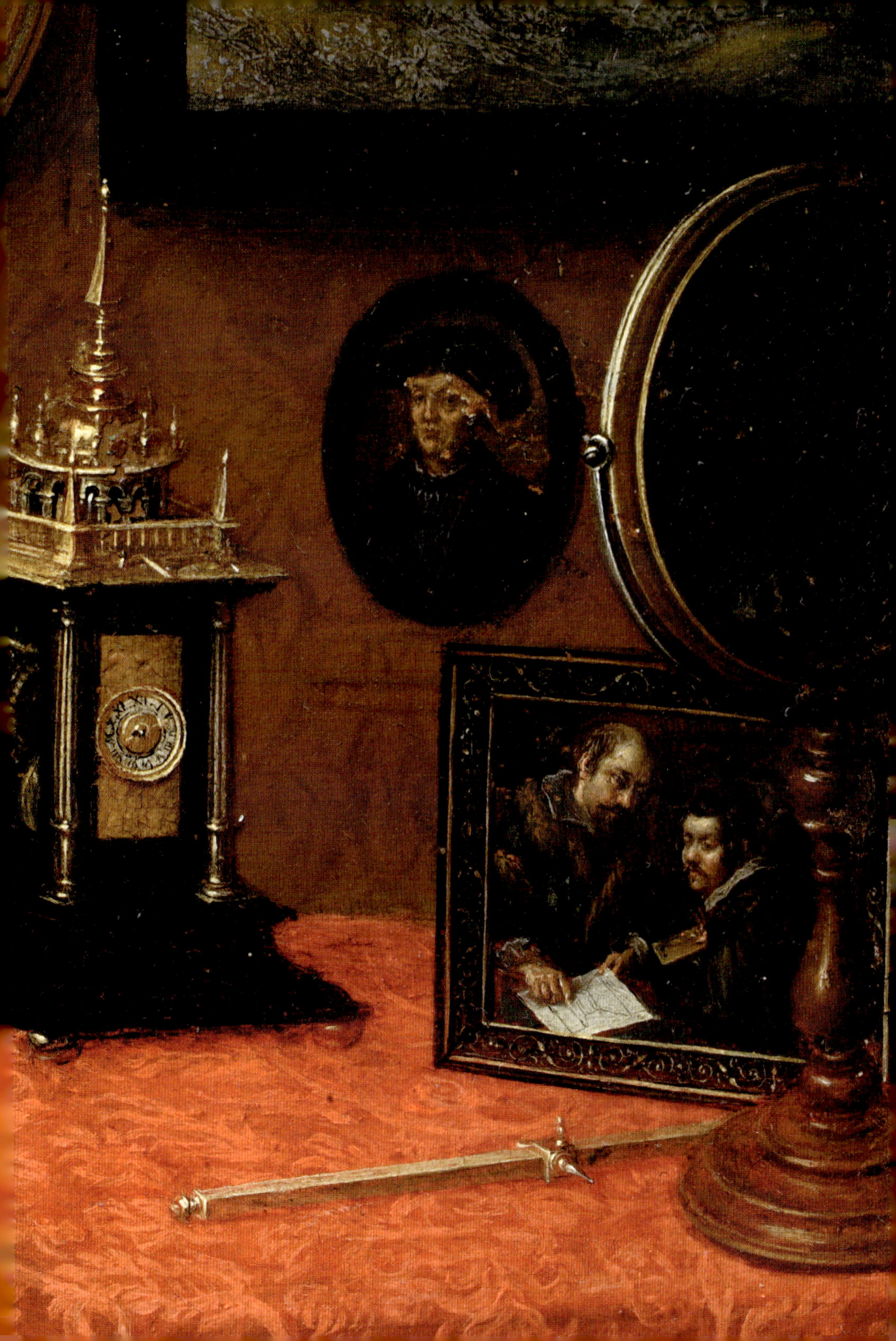

MARR | This double portrait and the mirror and the beam compass next to it relate to a portrait by another one of his pupils, the Milanese artist Daniele Crespi. Daniele Crespi was one of the most celebrated Milanese artists of the 1620s, he was a favorite of Cardinal Borromeo, the archbishop of Milan, who was also one of his patrons, and he painted this portrait, a double portrait, of Oddi teaching Peter Linder about the geometry of concave spherical mirrors. And in the Crespi double portrait we have the same mirror that we see in the bottom right-hand corner of the gallery interior and also the beam compass that rests next to it. And the composition of the double portrait in the gallery interior mimics that of Oddi and Linder, because Oddi and Linder are shown discussing a drawing. The drawing in the portrait of Oddi and Linder is in fact about the geometry of concave spherical mirrors. And in the gallery interior Oddi has been replaced by a painter and the diagram has been replaced by a perspective drawing of the room that we see in the gallery interior. So we know that there is a strong correlation between this image, painted in Milan in the

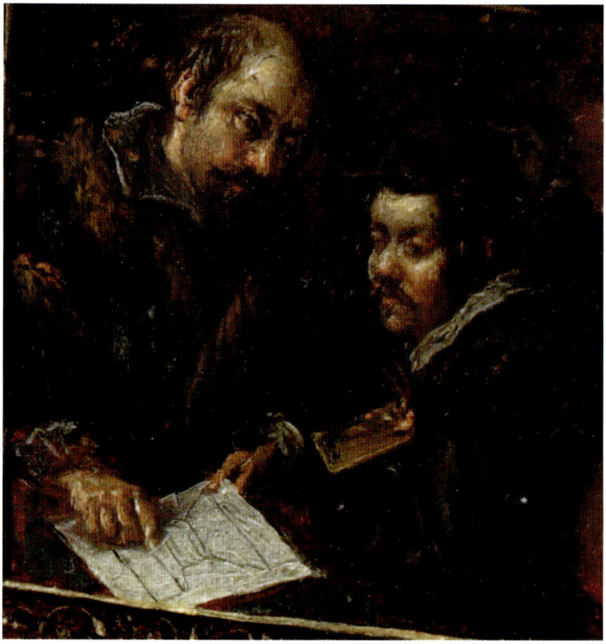

1620s, and this Flemish painting, the date of which must be around 1629 because that is when Caravaggio says he saw it in the house of Peter Linder. We have the coat of arms of Peter Linder in the top left-hand corner, and Linder is present in the gallery interior through his coat of arms and in his double portrait down here with the artist. Oddi, who, as Caravaggio tells us, was responsible in large measure for the design of the image, is present here in this portrait medal in the very centre of the table.

GORMAN | Which turned out to be the trickiest portrait medal to read in the picture.

CORDOVER | Yes, but the painting could easily have been done in the early 1620s although this letter was later. The only thing keeping this painting from being identifiable in the early 1620s is the issue of the *Rudolphine Tables* being published later, but we know that it was effectively completed ten years before.

GORMAN | The extraordinary thing is that this is a collaboration, an intellectual collaboration between at least three people. We have Linder, who is obviously

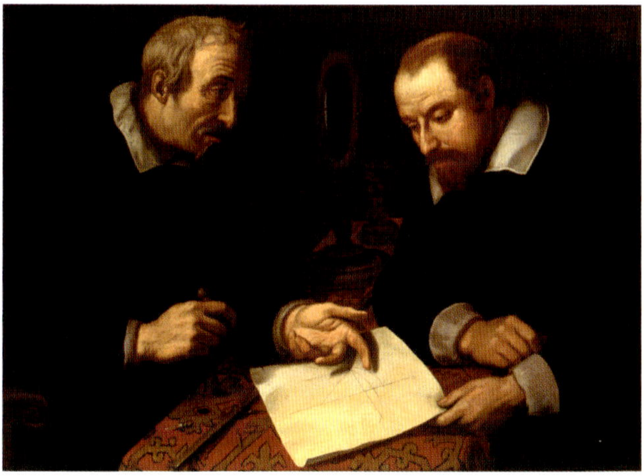

Daniele Crespi (Busto Arsizio/Milan 1597/1600–Milan 1630), attributed, *Portrait of Muzio Oddi and Peter Linder*, c. 1625, oil on canvas, 85 x 118 cm. Milan, Subert Collection.

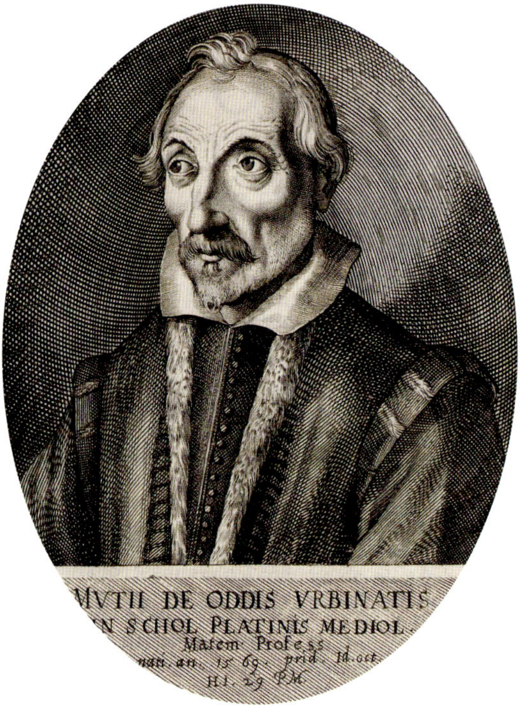

JOHANNES TROSCHEL, *Portrait of Muzio Oddi*, 1624, ENGRAVING. EDINBURGH, NATIONAL LIBRARY OF SCOTLAND, FIN. 85 (EXTRA PORTRAIT, IN MUZIO ODDI, *De gli horologi solari nelle superficie piane*, MILAN, GIACOMO LANTONI, 1614)

Medal portrait of Muzio Oddi, 1627, FROM PIER ANTONIO GAETANI, *Museum Mazzuchellianum, seu numismata virorum doctrina praestantium*, 2 VOLS., VENICE, ANTONIO ZATTA, 1761–3, II, PLATE CV, NO. IV. FLORENCE, BIBLIOTECA NAZIONALE CENTRALE DI FIRENZE, PALAT. 7.6.6.22.

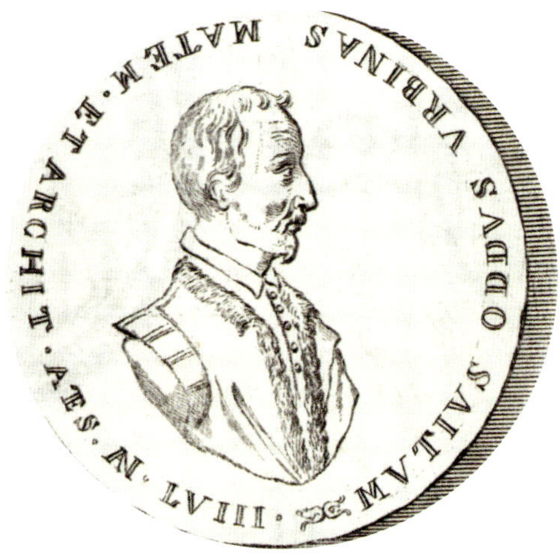

deeply involved as the patron in this project; we have Muzio Oddi as the mathematical instructor, and this is presumably some sort of lesson on the geometry of the burning mirror. And that is how we know that this is a concave mirror.

SMITH | And what is the date of this double portrait?

MARR | Mid-1620s.

GORMAN | So Linder is one of the richest people in Milan in the 1620s.

MARR | Muzio Oddi is basically the lynchpin. Oddi is Cardinal Borromeo's client, Oddi is the teacher of Linder and also of Borromeo's art agent, Ercole Bianchi. Bianchi is the man who travels between Antwerp and Milan for Cardinal Borromeo, bringing commissions, giving instructions to artists, including Jan Brueghel the Elder. So we've got this peculiar situation whereby we have a patron and his intellectual advisor in a sense—it's like the relationship between Borghini, Vasari and Cosimo de' Medici—in Milan, devising this picture which is being painted by—we presume, it's not certain—an artist either in or from Antwerp.

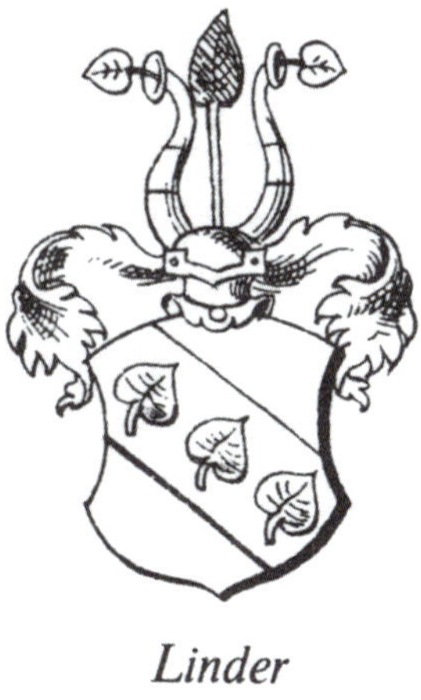

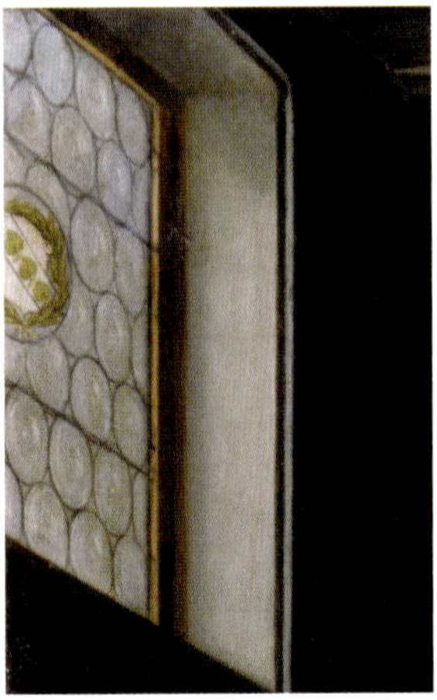

Linder coat of arms, from Ottfried Neubecker, *Großes Wappen-Bilder-Lexikon der bürgerlichen Geschlechter Deutschlands, Österreichs und der Schweiz*, Augsburg, 1997, p. 518.

GORMAN | And the interesting thing is that Jan Brueghel the Younger was in Milan in the 1620s.

MARR | And stayed at Bianchi's house.

GORMAN | And Bianchi is connected to Oddi, so it is very likely that there was a to and fro between Milan and Antwerp at the time through the various commissions that Borromeo was requesting from the Brueghel workshop, and it is also quite likely that Jan Brueghel the Younger played some role in this. It definitely links the picture to the Brueghel studio.

MARR | And Jan Brueghel the Elder said in a letter to Bianchi towards the end of his life that he had been planning a painting that depicted painting, sculpture and architecture, which of course is what this figure depicts. Now we don't know whether there is any causal connection, but this is clearly an idea that circulated within the Brueghel studio and was know to be out there.

GORMAN | I suppose a fundamental question remains: what is the relationship between the Windsor drawing and this painting? Because the drawing, I think, clearly is before the painting, and it's clearly very closely related to the picture—it doesn't seem to be a completely independent work. But the drawing, as you can see, is more conventional—it has the same octagonal table, it has the globe, but it has the more conventional picture of three connoisseurs conversing around a table, it has a dog... It is more in the traditional genre of gallery interiors.

CORDOVER | And I would argue that two of those cognoscenti are easily identifiable.

MARR | Yes.

CORDOVER | I think that the figure on the right is almost identical to a self-portrait by Peter Paul Rubens.

MARR | Yes, I would say that.

CORDOVER | And the figure on the left I think is also identifiable—

GORMAN | As Peter Linder?

MARR | Yes, I think it is Peter Linder.

CORDOVER | Do you think so?

GORMAN | Yes. And I also have a guess about who the figure in the middle is.

CORDOVER | That's extraordinary.

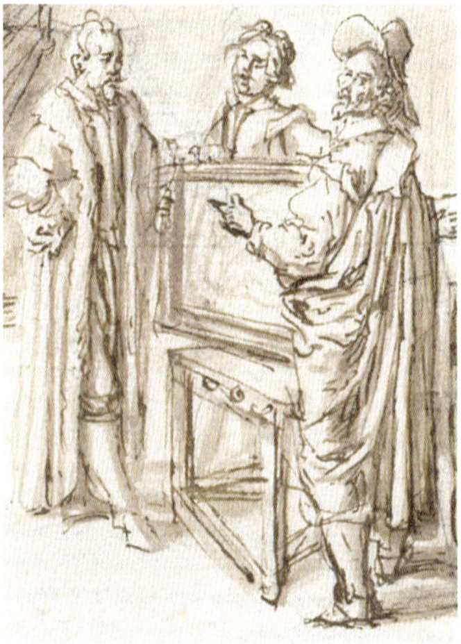

A Picture Gallery,
Windsor Castle, RL 12983,
detail.

Peter Paul Rubens
(after?), *Self-portrait*,
1623–5, oil on wood,
85 x 61 cm. Florence,
Uffizi, Vasari Corridor,
Inv. 1890, no. 1884.

GORMAN | I saw it this afternoon. I think it's Van Dyck, who was working in the Rubens studio from 1618 to 1620.

CORDOVER | Now that you've said it, it conjures up that portrait of Van Dyck with wild hair that does look just like this.

GORMAN | But if it is Van Dyck, think about the drawing—I think the drawing is memorializing a visit. Perhaps Linder went to Antwerp for business between 1618 and 1620, and this was a visit where he met Rubens and Van Dyck, and this is where the idea of him commissioning a gallery painting originated, and then this drawing was developed, possibly brought by Jan Brueghel the Younger to Milan. Then Linder and Oddi developed a more complex version of the composition and brought in the allegory, the deep mathematical content and the Kepler connection and so on, and then that led to the painting. It's just an idea.

CORDOVER | Van Dyck went to Milan with Jan Brueghel the Younger, his close friend, in 1622.

WESCHLER | The painting is painted in Antwerp or in Milan?

GORMAN | The painting could have been painted in theory in either, but it was definitely painted by an Antwerp painter and they didn't tend to stick around in Milan for long enough to do much, so it seems more likely that it was painted in Antwerp but commissioned perhaps on the basis of a visit to Milan.

WESCHLER | How could they have worked from Milan? There are the Alps in between.

SMITH | They did, though. Even Peter Brueghel.

MARR | It was very common, and we know this from the Brueghel correspondence, for a picture to be started in Antwerp, sent to Milan for approval, then sent back with a request for corrections, alterations, so we probably got a bit of copper shuttling backwards and forwards between Milan and Antwerp.

GORMAN | There are known examples of paintings that did shuttle back and forth in different states of completion. But there are still a lot of open questions around who created this picture... Was it just the work of one artist or was there more than one hand? Was it just that one figure with the goatee and the receding hairline in the double portrait, or was he just the person who completed the picture? I think it is fair to say that we are only at the beginning of understanding what's going on here.

CORDOVER | I want to ask Alex: who do you think painted the picture?

MARR | I think whoever painted the figure was whoever painted the rest of the picture, and there is no way that it was Van Dyck.

GORMAN | And who do you think it was? That's a very evasive statement.

MARR | I can say who it was not. There are very few people who painted gallery interiors who were active at this date. All of them, with the exception of Jan Brueghel, are ruled out on the basis of quality. Jan Brueghel is ruled out because he was dead.

SMITH | But what if the titles of the books were just added later?

CORDOVER | He's not dead if it was done in 1622 to 1624.

MARR | Yes, but there are elements of the painting that must have been done after that.

CORDOVER | There are only two elements, and the 1622–4 presence of each of them is easily explained. The *Rudolphine Tables* were known for ten years—they were finished well before the book was published. It didn't need to be published in order for it to be there, number one. And the medal of Oddi, if this is as we all agree a con-

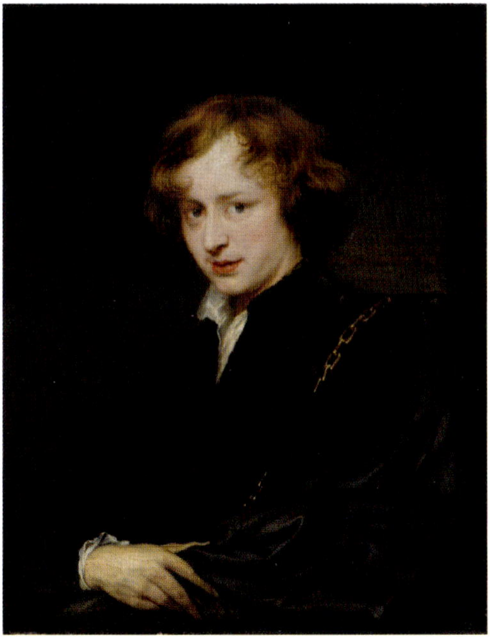

ANTHONY VAN DYCK (ANTWERP 1599–LONDON 1641), *Self-portrait*, *c.* 1622, OIL ON CANVAS, 82.5 X 70.2 CM. MUNICH, ALTE PINAKOTHEK.

ception of Oddi—how better to sign his work than to put a medal of his next to the medals portraying these great scientists and artists of the past? When I look very closely at this medal, it is slightly different from the 1627 one you've shown me.

GORMAN | It is always possible that these things were put in earlier, but I think you need a compelling reason to believe this, given that there are two elements of the picture that appear to be from 1627. Why would you add those things later? Why would you add the *Rudolphine Tables*, either add them after the painting was complete or include them before the book was published?

CORDOVER | I think they were included before the book was published; they are at the heart of the cosmology that is being described. 'Others may see it other ways.' It is a clarion call about the way the solar system works. The fact that the book was not yet published does not prevent me from thinking that it was well known—especially to Oddi. You asked, "What's the plausible reason for it to be completed prior to 1627?" My answer is quality.

MARR | Yes. There is no doubt in my mind that this is a work of exceptional quality. For me—you know, we have been backwards and forwards on this many, many times—for me, stylistically, it is closest to the *Cognoscenti* painting in the National Gallery in London in terms of palette and manner.

CORDOVER | Agreed.

MARR | And I think that work and this work, while they share similarities to Brueghel's manner—I would say whoever painted this spent time in Brueghel's studio. Stylistically, it is just sufficiently dissimilar to suggest that we shouldn't attribute it to Jan Brueghel the Elder. It is certainly not Jan Brueghel the Younger, it's way too good.

GORMAN | Though parts of it could be Jan Brueghel the Younger. He was in the right place at the right time.

MARR | Parts of it could be—I mean, if there were some additions. The clumsy lettering, for example, that could be. So we shouldn't leave out the possibility that there were a few later additions. But we've got two works of exceptional quality, both of which are by painters from the Brueghel milieu who are currently unidentified. There are oodles of unattributed paintings around the world—that is not a barrier at all.

GORMAN | I think it is the same principal artist who created both the Windsor drawing and the painting.

MARR | The hand in the Windsor drawing is totally different to the preparatory drawing for the London picture. That doesn't necessarily mean anything, but it is worth noting. And it is also worth noting that these drawings circulated very freely between the studios, so whoever drew this—while I think that it is probably likely that it was the person who painted the *Linder Gallery*—might not have been involved in the development of the drawing into a painting at all.

GORMAN | I think it is compelling because of the sketches.

CORDOVER | I agree.

MARR | Yep. I agree, but let's just keep it open as a possibility. I am not saying that it is, but it's possible. It is possible that this is a drawing by a different artist to whoever painted this.

WESCHLER | You know what I think? I think different people see it different ways.

PRESENTLY UNATTRIBUTED
(POSSIBLY JAN BRUEGEL THE ELDER AND/OR HIS STUDIO)
Linder Gallery
c. 1622–9
oil on copper, 56.5 x 82.2 cm
New York, Cordover Collection

Provenance: collection of Baron Anselm von Rothschild, Vienna; collection of Baron Nathaniel von Rothschild, Vienna; collection of Baron Alfons von Rothschild, Vienna; collection of Baroness Clarice Sebag-Montefiore, Vienna; Newhouse Galleries, New York; collection of Thomas Mellon Evans, prior to acquisition by the present owner.

The painting shows the interior of a vaulted gallery in perspective, with a balustrade looking out over an Italianate garden containing a fountain and an obelisk. The walls of the room are hung with paintings in the styles of Flemish and Italian artists. Sculptural works, including both copies of classical sculptures and recent works such as Giambologna's *Bull*, books and musical instruments are also present in the gallery. Three tables bear an impressive array of mathematical instruments, drawings and diagrams.

The central octagonal table includes a celestial globe, a large Arsenius astrolabe, a Jacob's staff, a Galileian geometric and military compass (with an alidade not found in Galileo's original instrument) and a perspective-drawing instrument. This table also displays a variety of compasses, pens, an inkwell, lenses, an hourglass, a book of drawings and an engraving that can be identified as the *Martyrdom of St Catherine* by the Master MZ. The books to the right of the celestial globe can be identified as *The Rudolphine Tables* (1627) and *The Harmony of the World* (1619), both by German astronomer Johannes Kepler (1571–1630), and the *Description of the Admirable Table of Logarithms* (1614) by Scottish mathematician John Napier (1550–1617). At the front of the table is a diagram depicting three competing cosmic systems: the Ptolemaic system, the Copernican system and the 'compromise' Tychonic system in which the inner planets orbit the Sun and the remaining planets orbit the Earth. Beneath the three systems is the inscription ALY ET ALIA VIDENT, 'Different people see it differently' or 'Others see it yet otherwise'. Partially hidden behind the diagram of the cosmic systems is an astrological geniture. Also prominently displayed on the table are identifiable portrait medals of Muzio Oddi (1569–1639), Girolamo Cardano (1501–76), Andrea Alciati (1492–1550), Albrecht Dürer (1471–1528), Michelangelo Buonarroti (1475–1564) and Donato Bramante (1444–1514).

The allegorical figures in the *Linder Gallery* are of particular interest. The bearded male figure can probably be identified as Disegno or Drawing, and may be intended as a likeness of astronomer Johannes Kepler, whose books are represented prominently on the octagonal table. The reclining female figure in classical dress, with paintbrushes, mallet, maulstick and a book, and wearing a laurel wreath and a sun pendant, may be identified as a personification of Pictura (Painting), or perhaps more broadly 'the Arts'. The laurel wreath and sun pendant are the attributes of Virtue, according to Cesare Ripa's *Iconologia*, so the allegory can probably be read as 'The Arts and Virtue rest on Design' (punning on the double meaning of design as 'drawing' and 'purpose').

The identity of the patron of the painting may be deduced from the coat of arms in the upper left-hand window showing three leaves in a diagonal band, the arms of the Linder family. In a letter sent to Urbino mathematician and architect Muzio Oddi, Oddi's friend and pupil Giovan Battista Caravaggio described seeing this painting in 1629 during a visit to the house of German merchant Peter Linder in Milan. According to Caravaggio, Oddi was himself largely responsible for the 'invention' of the painting; this is further supported by the presence of a portrait medal of Oddi on the central octagonal table. A small double portrait on the right-hand red table depicts the patron, identified as Linder, pointing at a drawing. On close inspection the drawing may be seen to be a perspective scheme for the whole painting. Another figure, presumably the artist, is looking at the perspective drawing and painting from it. A preparatory drawing for the *Linder Gallery* has been discovered in Windsor Castle, with a group of conversing connoisseurs replacing the allegorical figures. An identification of the artist of this remarkable painting remains challenging. Jan Brueghel the Younger visited Milan in the early 1620s, when the painting is likely to have been commissioned, but stylistically the painting appears more consistent with the collaborations between Jan Brueghel the Elder (who died in 1625, before the painting is likely to have been completed) and Hendrick van Balen.

MICHAEL JOHN GORMAN

ALEXANDER MARR

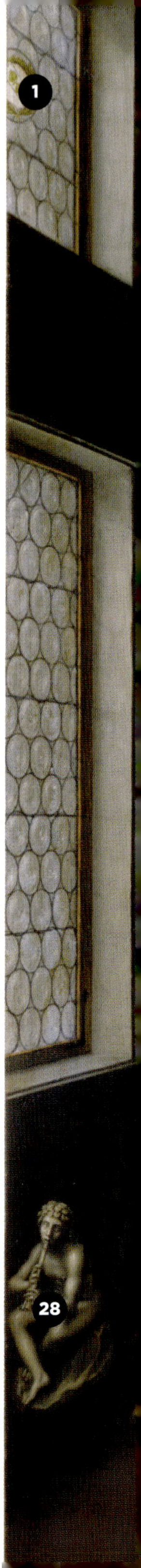

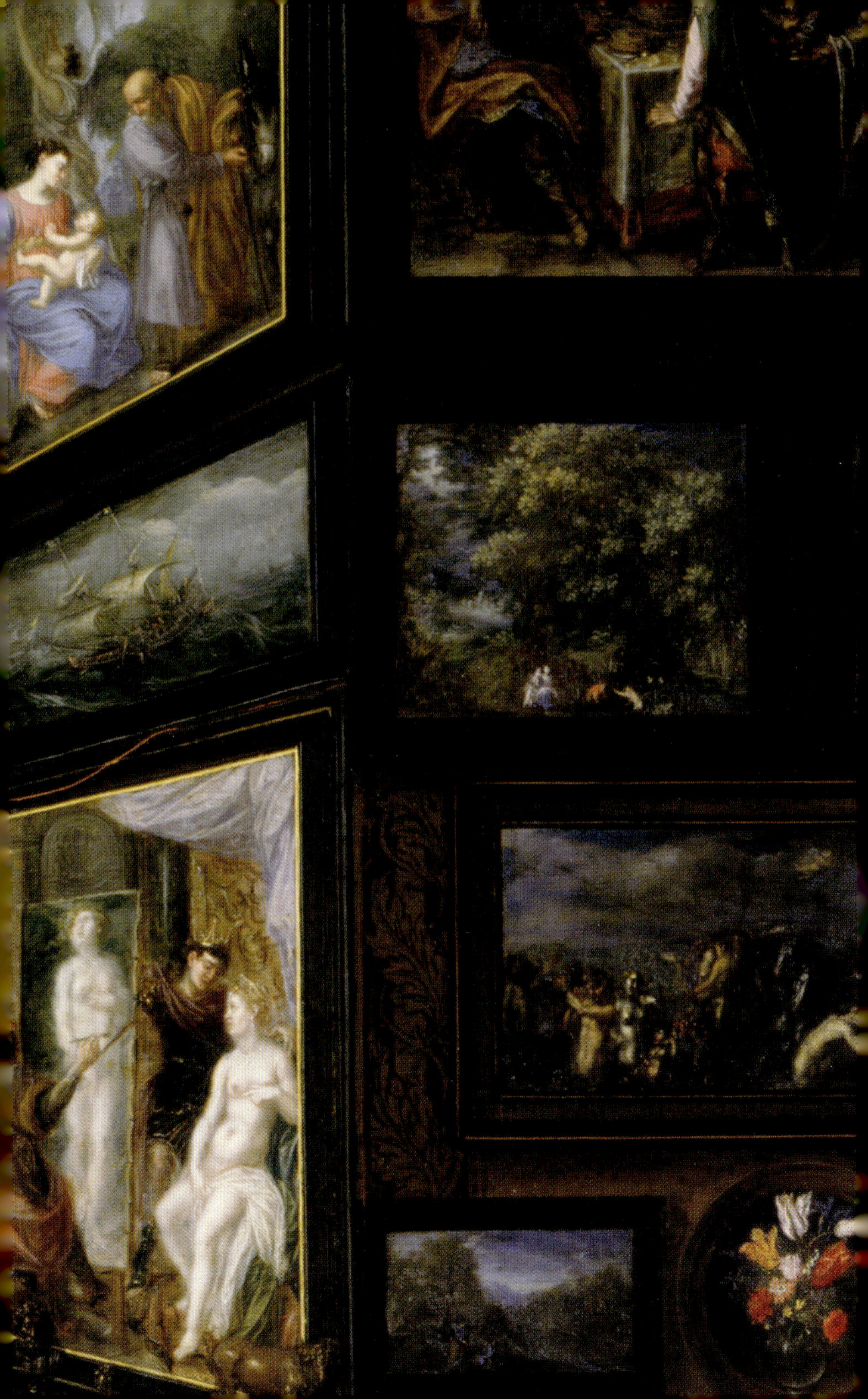

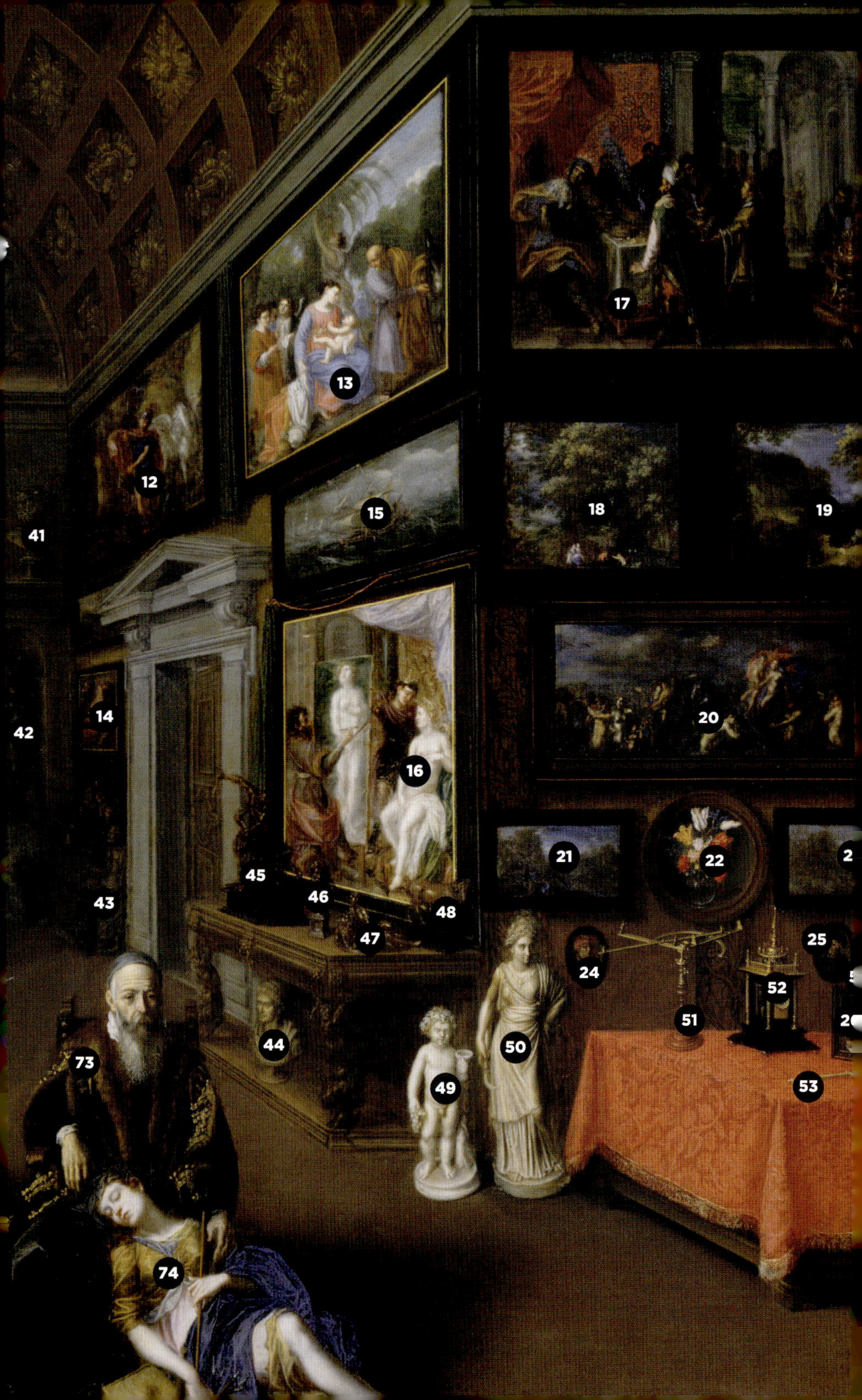
13
17
12
15
18
19
41
14
42
20
16
21
22
45
46
43
48
47
25
24
52
51
44
50
73
49
53
74

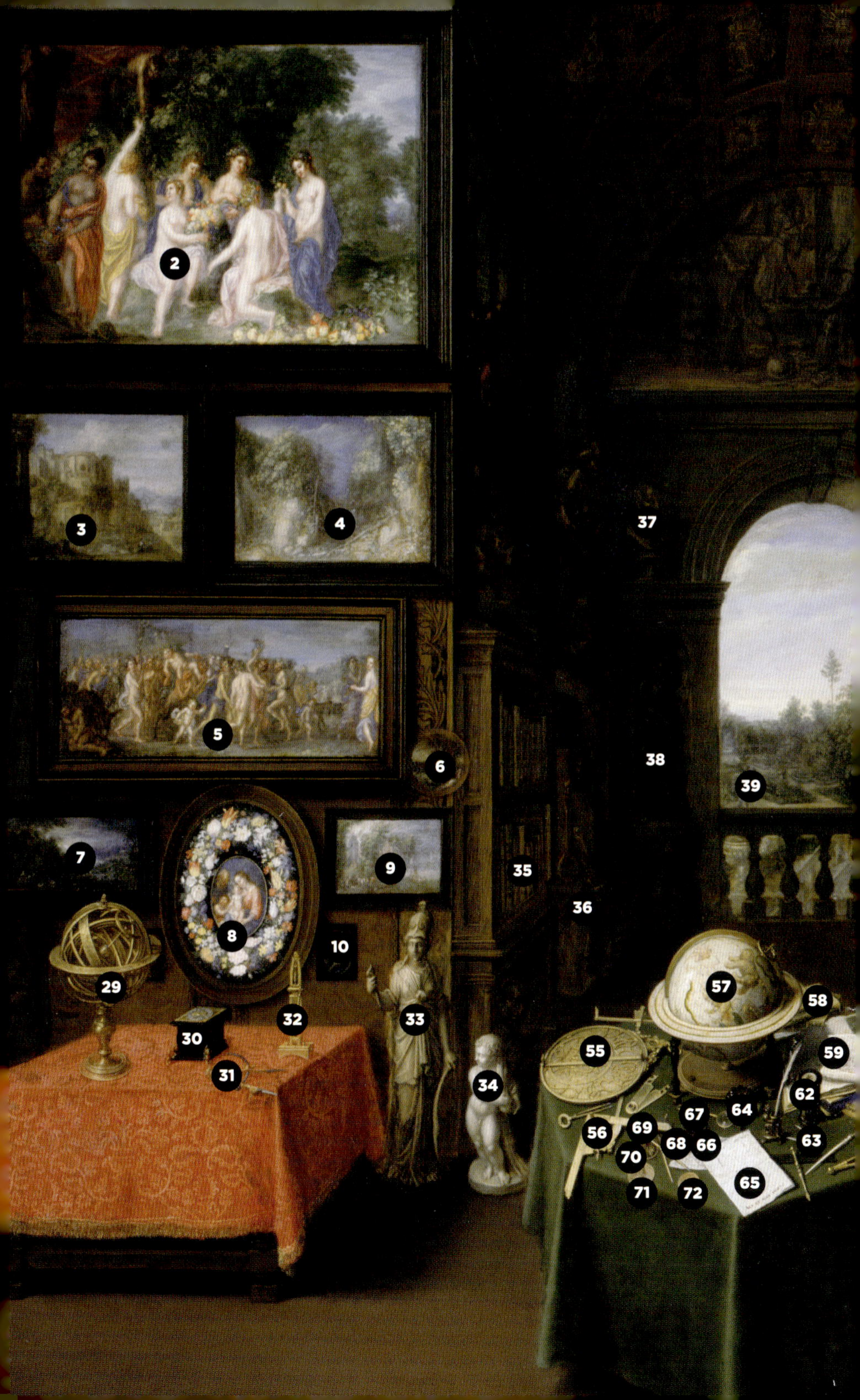
2
3
4
37
5
6
38
39
7
35
9
36
8
10
29
57
58
32
33
30
55
59
31
34
62
64
67
56
69
68
66
63
70
65
71
72

1| Linder coat of arms. This coat of arms, with three leaves in a diagonal band, is the arms of the Linder family, suggesting the identity of the commissioner of the work as Nuremberg merchant Peter Linder, active in Milan in the 1620s before moving to Venice, where he became *console* of the Fondaco dei Tedeschi.

2| *Nymphs and Satyrs* in the style of Hendrick van Balen (Antwerp, 1575–1632).

3| *View of Rome with the Palatine Hill* in the style of Willem van Nieulandt the Younger (Antwerp, 1584–Amsterdam, 1635/6).

4| *Landscape* in the style of Gillis van Coninxloo (Antwerp, 1544–Amsterdam, 1607).

5| *Triumph of Bacchus* in the style of Frans Francken II (Antwerp, 1581–1642).

6| *Landscape with Figures.*

7| *Nocturnal Landscape* in the style of Lucas van Uden (Antwerp, 1595–1672).

8| *The Holy Family with St John in a Flower Garland* in the style of Jan Brueghel the Elder (Brussels, 1568–Antwerp, 1625) and Hendrik van Balen (Antwerp, 1575–1632).

9| *River Landscape* in the style of Jan Brueghel the Elder (Brussels, 1568–Antwerp, 1625).

10| *Portrait Miniature of Albrecht Dürer* (Nuremberg, 1471–1528).

11| *Ignorance Destroying the Arts and Sciences* in the style of Frans Francken II (Antwerp, 1581–1642).

12| *Perseus with the Head of Medusa* in the style of Jan Boeckhorst (Münster, *c.* 1604–Antwerp, 1668).

13| *The Rest on the Flight to Egypt with the Miracle of the Leaning Palm* in the style of Carlo Saraceni (Venice, *c.* 1579–1620).

14| *The Death of Lucretia.*

15| *Seascape* in the style of Andries van Eertvelt (Antwerp, 1590–1652).

16| *Apelles Painting Campaspe*, in the style of Joos van Winghe (Brussels, *c.* 1544–Frankfurt, 1603).

17| *Belshazzar's Feast* in the style of Kasper van den Hoecke (1595–Antwerp, after 1648).

18| *Landscape with Figures* in the style of Abraham Govaerts (Antwerp, 1589–1626).

19| *The Flight into Egypt* (?) in the style of Paul Bril (Antwerp, 1554–Rome, 1626).

20| *Triumph of Neptune and Amphitrite* in the style of Frans Francken II (Antwerp, 1581–1642).

21| *Landscape with Tobias and the Archangel Raphael Returning with the Fish.*

22| *Still Life of Flowers in a Vase* in the style of Osias Beert the Elder (Antwerp, *c.* 1580–1624).

23| *Extensive Landscape.*

24| *Portrait miniature of Jan van Eyck* (Maaseik, before *c.* 1395–Bruges, 1441).

25| *Portrait miniature of Lucas van Leyden* (Leiden, 1494–1533).

26| *Double Portrait of Peter Linder and an Unidentified Artist.* This is likely a self-portrait of the artist with his patron indicating a perspective scheme for the painting. The double portrait is similar to a double portrait of Muzio Oddi and Peter Linder by Daniele Crespi (see p. 115).

27| Statue of Apollo.

28| Statue of Pan (?).

29| Armillary sphere (German, 16th century).

30| Ebony and gilt table clock (probably Augsburg, 16th century).

31| Pair of compasses.

32| Gunner's level, possibly by Ulrich Schniep (Wiesensteig, active from 1545–d. 1588).

33| Statue of Minerva.

34| Statue of Amor (?).

35| Bookcase including a book by Euclid.

36| Musical instruments. Along with astronomy, geometry and arithmetic, music was part of the *quadrivium*, which combined with the *trivium* of logic, rhetoric and grammar made up the seven liberal arts.

37| Bust of Julius Caesar (?).

38| Statue of Adam (?).

39| Fountain with putto playing the bagpipes. This fountain bears a resemblance to the Singing Fountain in front of the Belvedere in Prague, where Kepler and Tycho Brahe carried out observations. The fountain could also represent hydraulics as part of mathematics.

40| Obelisk. While this is an unlikely architectural feature, the obelisk had connections with astronomical measurement in the seventeenth century, being associated particularly with solar observation.

41| Bust of Hercules (?).

42| Statue of Eve (?).

43| *Perpetuum mobile* by Cornelis Drebbel (Alkmaar, 1572–London, 1633). Powered by changes in air-pressure, this device (only barely visible) became well-known around 1612 and was a device of fascination to Flemish artists including Rubens.

44| Bust of Nero.

45| *Hercules and a Centaur* by Giambologna (Jean Boulogne; Douai, 1529–Florence, 1608).

46| Bust of Septimius Severus (?).

47| Sculpture of Cybele (?) and a lion.

48| *Bull* by Giambologna (Jean Boulogne; Douai, 1529–Florence, 1608).

49| Statue of Bacchus.

50| Statue of Ceres.

51| Triangulation instrument for military surveying.

52| Ebony and gilt German clock, *c.* 1600.

53| Beam compass, used to describe long-radius circles.

54| Concave ('burning') mirror viewed from the back. This mirror is also found in a double portrait of Muzio Oddi and Peter Linder painted by Daniele Crespi.

55| Large astrolabe by Gualterus Arsenius (Walter Aertsen; fl. 1556–75), similar to the 'Philip II Astrolabe' in Madrid.

56| Galileian compass with alidade. This instrument may be a version of Galileo's astrolabe created by the Barocci–Vagnarelli workshop in Urbino.

57| Celestial globe, probably by Jodocus Hondius the Younger (Amsterdam, 1593–1629).

58| Two books by Johannes Kepler (Weil der Stadt, 1571–Regensburg, 1630) are shown here: *Tabulae Rudolphinae* (1627) and *Harmonices Mundi* (1619). The book on top is the *Mirifici logarithmorum canonis descriptio* by John

Napier (Merchiston Castle, 1550–Edinburgh, 1617).

59| Engraving of the *Martyrdom of St Catherine* in the style of the Master MZ.

60| Drawing book.

61| Perspective instrument in the style of those by Wenzel Jamnitzer (Vienna, 1507/8–Nuremberg, 1585) or Ludovico Cigoli (San Miniato, 1559–Rome, 1613), with a perspective drawing of a colonnade.

62| Hourglass.

63| Compasses and drawing implements in the style of the Barocci–Vagnarelli workshop in Urbino.

64| Convex and concave lenses.

65| Drawing of the three cosmic systems: the Earth-centred Ptolemaic system, the Sun-centred Copernican system and the compromise Tychonic system, with the inscription ALY ET ALIA VIDENT, 'Others see it yet otherwise'.

66| Astrological geniture of an unidentified person born in the month of March.

67| Portrait medal of Muzio Oddi (Urbino, 1569–1639) by an unknown author, 1627. Inscription: MVTIVS ODDVS VRBINAS MATEM(aticus) ET ARCHIT(ectus) AE(tatis) S(uae) AN(ni) LVIII.

68| Portrait medal of Andrea Alciati (Alzate Brianza/Milan, 1492–Pavia, 1550), Milanese legal scholar and author of the *Emblemata*, by Jean Second (Jan Everaerts; The Hague, 1511–Saint-Amand, Fleurus, 1536). Inscription: ANDR(eas) ALCIATVS IVRECO(n)S(ultus) COMES P(alatinus).

69| Portrait medal of Girolamo Cardano (Pavia, 1501–Rome, 1576), Milanese physician, astrologer and philosopher, attributed to Leone Leoni (Arezzo, 1509–Milan, 1590), *c.* 1550. Inscription: HIER(onymus)·CARDANVS·AETATIS·AN(norum)·XLVIIII.

70| Portrait medal of Albrecht Dürer (Nuremberg, 1471–1528) by Mathes Gebel, executed in 1528, upon Dürer's death. Inscription on the recto: IMAGO·ALBERTI·DVRERI·AETATIS·SVAE·LVI.

71| Portrait medal of Michelangelo Buonarroti (Caprese, 1475–Rome, 1564) by Leone Leoni (Arezzo, 1509–Milan, 1590). Inscription: MICHAELANGELVS·BONARROTVS·FLOR(entinus)·AET(atis)·S(uae)·ANN(orum) 88·.

72| Portrait medal of Donato Bramante (Monte Asdrualdo, now Fermignano, Pesaro, 1444–Rome, 1514), architect. Inscription: BRAMANTES·

ASDRVVALDINVS. Muzio Oddi is known to have owned a copy of this medal which has previously been attributed to Caradosso (Cristoforo Caradosso Foppa) but is now thought to be a self-portrait.

73| The bearded male figure is probably intended as an allegorical representation of Disegno (drawing or design), but could also be a portrait. It is possible that this is a likeness of German astronomer Johannes Kepler (1571–1630).

74| The female allegorical figure in classical dress with palette and brushes, maulstick, mallet and a book appears to represent Painting, Sculpture and possibly Architecture. The laurel wreath and sun pendant indicate that she also represents Virtue. She appears to be resting on the male figure, suggesting the interpretation 'The Arts and Virtue rest on Design', where Design encompasses both mathematics and astronomy.

SELECTED BIBLIOGRAPHY

Belting, Hans. *Likeness and Presence: A History of the Image before the Era of Art*, Chicago, 1997 (2nd German edition, Munich, 1991).

Caspar, Max. *Kepler*, translated and edited by Clarice Doris Hellman, with an introduction and references by Owen Gingerich, New York, 1993 (originally published London–New York, 1959; first German edition, Munich, 1948).

Filipczak, Zirca Zaremba. *Picturing Art in Antwerp, 1550–1700*, Princeton, N.J., 1987.

Freedberg, David. *The Power of Images: Studies in the History and Theory of Response*, Chicago, 1989.

Gorman, Michael John; Marr, Alexander. "'Others see it yet otherwise': disegno and pictura in a Flemish gallery interior", *The Burlington Magazine*, no. 1247, 149, 2007, 85–91.

Härting, Ursula. "Doctrina et Pietas: über frühe Galeriebilder", *Jaarboek Koninklijk Museum voor Schone Kunsten Antwerpen*, 1993, 95–133.

Honig, Elizabeth A. *Painting and the Market in Early Modern Antwerp*, New Haven, Conn., 1998.

Important Old Master Paintings from the Thomas Mellon Evans Collection, auction catalogue (Christie's New York, Friday, 22 May 1998), Lot 8.

Speth-Holterhoff, Simone. *Les Peintres Flamands de Cabinets d'Amateurs au XVIIe Siècle*, Brussels, 1957.

Stoichita, Victor I. The *Self-Aware Image: An Insight into Early Modern Meta-Painting*, Cambridge, 1997.

The Age of Rubens, exhibition catalogue (Boston, Mass., 1993–4; Toledo, Oh., Toledo Museum of Art, 1994), edited by Peter C. Sutton, with the collaboration of Marjorie E. Wieseman, Boston, 1993.

Winner, Matthias. *Die Quellen der Pictura-Allegorien in gemalten Bildergalerien des 17. Jahrhunderts zu Antwerpen*, PhD Diss., Cologne, Universität Köln, 1957.